THE POLITICS OF FISHING

Also by Tim S. Gray

BURKE'S DRAMATIC THEORY OF POLITICS (*with Paul Hindson*)

FREEDOM

THE FEMINISM OF FLORA TRISTAN (*with Maire Cross*)

THE POLITICAL PHILOSOPHY OF HERBERT SPENCER:
Individualism and Organicism

UK ENVIRONMENTAL POLICY IN THE 1990s

The Politics of Fishing

Edited by

Tim S. Gray
Professor of Political Thought
University of Newcastle

First published in Great Britain 1998 by
MACMILLAN PRESS LTD
Houndmills, Basingstoke, Hampshire RG21 6XS and London
Companies and representatives throughout the world

A catalogue record for this book is available from the British Library.

ISBN 0–333–69085–0

First published in the United States of America 1998 by
ST. MARTIN'S PRESS, INC.,
Scholarly and Reference Division,
175 Fifth Avenue, New York, N.Y. 10010

ISBN 0–312–21410–3

Library of Congress Cataloging-in-Publication Data
The politics of fishing / edited by Tim S. Gray.
p. cm.
Includes bibliographical references and index.
ISBN 0–312–21410–3 (cloth)
1. Fishery policy. 2. Fishery management. 3. Fishery policy–
–European Union countries.   4. Fishery management—European Union
countries.   I. Gray, Tim, 1942–   .
SH328.P63 1998
338.3'727'094—DC21   98–10620
                                                                                    CIP

This book is printed on paper suitable for recycling and made from fully managed and
sustained forest sources.

10   9   8   7   6   5   4   3   2   1
07 06 05 04 03 02 01 00 99 98

Printed and bound in Great Britain by
Antony Rowe Ltd, Chippenham, Wiltshire

To the memory of the eight hundred UK fishers whose lives have been lost at sea during the last forty years

Contents

Acknowledgements

This is to acknowledge with gratitude the generous support given by the following sources to support the colloquium on the politics of fishing held at the University of Newcastle on 17–18 September 1996, from which this book originates:

Economic and Social Research Council's Global Environmental Change Programme Network and Dissemination Fund

Humanities Research Board of the British Academy

Newcastle University Research Committee

Notes on the Contributors

Charles Cann retired from the Civil Service in 1996, having been from 1991 to 1996 the Deputy Secretary in the Ministry of Agriculture, Fisheries and Food (MAFF) responsible, among other things, for the Ministry's Fisheries Department. He was involved in EU and UK fisheries policy for 15 of his last 20 years with MAFF.

Kevin Crean has extensive experience of the planning, management and development of fisheries in the UK, Africa, Europe (including Eastern Europe and Russia) and the Pacific Rim countries. His publications reflect this generalist background but more recent work has focused on fishermen's organizations in the UK and the development of alternative approaches to coastal fisheries management. He holds the post of Deputy Director at the University of Hull International Fisheries Institute.

Barrie Deas is the Chief Executive of the National Federation of Fishermen's Organizations, the representative body for fishermen in England, Wales, Northern Ireland and the Channel Islands. In addition to its primary role as a trade association promoting its members' interests at local, national and EU level, the NFFO has been at the forefront of developing a viable alternative to the current Common Fisheries Policy. The NFFO's policy papers include 'Conservation: An Alternative Approach' (1993); 'Coastal State Management: An Alternative to the Common Fisheries Policy' (December 1996); and 'Coastal State Management: A Strategy for Implementation' (May 1996).

John Goodlad was born and brought up in the Shetland fishing village of Hamnavoe and graduated from Aberdeen University in 1978. After obtaining a postgraduate degree for fisheries research, he was employed as Fisheries Development Officer for the Shetland Islands Council. Since 1982 he has been employed as Chief Executive of both the Shetland Fishermen's Association and the Shetland Fish Producers' Organization. In these capacities he has been closely involved in the expansion and development of the Shetland seafood industry during the past 15 years. A speaker at various international conferences, he has published a variety of papers in several fishing journals and periodicals. He is a board member of Highlands and Islands Enterprise and was chairman of the CFP Review Group which was established by Fisheries Ministers in 1995.

Mark Gray is a fisheries researcher based in Anglesey. He holds an MSc in Fisheries Biology and Management from Bangor University, and has been employed by MAFF, the Environment Agency and WWF (UK) on research contracts connected with coastal and fresh-water fishing. He has also served as a Senior EU Fisheries Observer on Portuguese vessels fishing off the Canadian Grand Banks. His publications include *The Coastal Fisheries of England and Wales, Part III: A Review of the Status 1992–1994*, Fisheries Research Technical Report No. 100 (Lowestoft: MAFF Directorate of Fisheries Research, 1995).

Tim S. Gray is Professor of Political Thought and Head of the Department of Politics at the University of Newcastle. He has published several books, articles and chapters on political thought together with several chapters on environmental policy, and has edited *UK Environmental Policy in the 1990s* (Macmillan, 1995). He is currently working on a study of the impact of the environmental movement on fishing.

Brendan McNamara is a consultant in European business affairs and was for many years a senior official in the European Commission. His publications include the first periodic report on the regions of Europe as well as several studies on development aid under the Lomé Conventions. He has also prepared a study on rural development for the OECD and a report, *Crusade for Survival*, on development of the West of Ireland.

Knut H. Mikalsen is Associate Professor, Department of Political Science, University of Tromsø. His recent publications include 'Lessons from the Abyss: Reflections on Recent Fisheries Crises in Atlantic Canada and Norway', *Dalhousie Law Journal*, 18(1), 1995 (with R. Apostle), and 'Regulating Fjord Fisheries: Folk Management or Interest Group Politics?' in C. L. Dyer and J. R. McGoodwin (eds), *Folk Management in the World's Fisheries* (University Press of Colorado, 1994, with S. Jentoft).

Robert O'Connor, who died in January 1997, was Emeritus Professor in the Economic and Social Research Institute, Dublin and a leading figure in the analysis of the development needs of the Irish economy. His work covered a wide range of topics relating to the economics of natural resource use. He made a number of important contributions to fisheries research, notably (with others) in *Development of the Irish Sea Fishing Industry and its Regional Implications* (1980) and, again with other authors, in *Review of the Irish Aquaculture Sector* (1992), both published by the ESRI.

Tim Oliver has been editor of *Fishing News* for ten years and was formerly a fisherman in the distant-water trawling industry which was all but wiped out by the imposition of 200-mile limits by Iceland, Norway and Canada in the 1970s. *Fishing News* is a weekly trade newspaper covering all aspects of the commercial fishing industry which is universally read by all sectors of the industry.

Gísli Pálsson (PhD, Manchester University, 1982) is Professor of Anthropology, University of Iceland, Reykjavik, and (formerly) Research Fellow at the Swedish Collegium for Advanced Study in the Social Sciences, Uppsala, Sweden. He has published a number of articles on fishing, practical skills and environmental issues in international journals. Among his books are *The Textual Life of Savants: Ethnography, Iceland and the Linguistic Turn* (Harwood Academic Publishers, 1995) and *Nature and Society; Anthropological Perspectives* (co-edited with P. Descola, Routledge, 1996).

Jeremy Phillipson is currently working within the School of Geography and Earth Resources at the University of Hull, where he is Network Manager of the European Social Science Fisheries Network (FAIR CT95 0070). His publications include 'The Sustainable Development of UK Fisheries: Opportunities for Co-management', *Sociologia Ruralis*, 36(2), 1996, and 'The Imperative of Institutional Reform: Alternative Models and the UK Fishing Industry', in the *Proceedings of the VIIth Annual Conference of the European Association of Fisheries Economists*, 10–12 April 1995 (CEMARE: University of Portsmouth, 1996).

David Steel has been involved in fisheries policy for over twenty years. He joined the White Fish Authority in 1976 as a fisheries economist with the Fisheries Economics Research Unit. He produced a wide variety of research reports, concentrating on the structural aspects of the British industry and those of other European fishing industries. He participated in consultancy projects in the UK and overseas and was also editor of *Fisheries Economics Newsletter*. Since 1981 he has been employed in the European Parliament, where he has worked on fisheries, agriculture and budgetary questions. In recent years his papers on the Common Fisheries Policy, including structural questions, the regulation of fisheries and, in particular, the role of the European Parliament, have been published in *Marine Policy* by the Council of Europe and by the International Institute of Fisheries, Economics and Trade.

David Symes is a Senior Lecturer in Geography at the University of Hull and has gained international recognition in the fields of both rural sociology and fisheries management. He has recently led an EU-funded, six-nation research project on Devolved and Regional Management Systems in Fisheries and is currently coordinating the European Social Science Fisheries Network (ESSFiN) also funded by the EU as a Concerted Action. He is the author of well over one hundred academic papers and has recently co-edited *Fisheries Management in Crisis* (Blackwell Science, 1996) and a Special Issue of *Sociologia Ruralis*, 36(2), 1996 on 'Sustainable Fisheries?'.

Mireille Thom is a researcher based in the Highlands of Scotland. Her main research interests include the implementation of European policies and European funds and their effects on rural areas with the Common Fisheries Policy as the main focus. Since completing her PhD thesis, entitled *The Governance of a Common in the European Community: The Common Fisheries Policy* (1993), she has carried out several research projects for various organizations in the UK and abroad. She also contributes to several maritime publications in the UK and France.

List of Abbreviations

ACF	Advisory Committee on Fisheries
ACFM	Advisory Committee on Fisheries Management
ASFC	Association of Sea Fisheries Committees
BTF	British Trawler Federation
CCPM	Comité Central des Pêches Maritimes
CFP	Common Fisheries Policy
CFPRG	Common Fisheries Policy Review Group
CLPM	Comités Local des Pêches Maritimes
CNPMCM	Comité National des Pêches Maritimes et des Cultures Marines
CPRs	Common Pool Resources
DANI	Department of Agriculture, for Northern Ireland
DF	Danmarks Fiskeriforening (Danish Fishermen's Association)
DFO	Department of Fisheries and Oceans
DG XIV	Fisheries Directorate
ECJ	European Court of Justice
EEA	European Economic Area
EEZ	Exclusive economic zone
EFTA	European Free Trade Association
EP	European Parliament
ERDF	European Regional Development Fund
ERM	Exchange Rate Mechanism
ESF	European Social Fund
EU	European Union
FA	Fishermen's Association
FAL	Fishermen's Association Limited
FAO	Food and Agriculture Organization of the United Nations
FCA	Fisheries Cooperative Association
FCG	Fisheries Conservation Group
FHIF	Federation of the Highlands and Islands Fishermen
FIFG	Financial Instrument for Fisheries Guidance
FO	Fishermen's Organization
FPO	Fish Producer Organization (same as PO)
GRT	Gross registered tonnage

ICES	International Council for the Exploration of the Sea
ICNAF	International Commission for Northwest Atlantic Fisheries
IGC	Intergovernmental Conference
IMM (97)	Intermediate Ministerial Meeting on the Integration of Fisheries and Environmental Issues in 1997
IQ	Individual Quota
ISWFO	Irish South and West Fishermen's Organization
ITQ	Individual Transferable Quota
IVQ	Individual Vessel Quota
LDC	London Dumping Convention
LPO	Local producer organization
MAFF	Ministry of Agriculture, Fisheries and Food
MAGP	Multi-Annual Guidance Programme
MLS	Minimum Landing Size
MMS	Minimum Mesh Size
MSA	Merchant Shipping Act
NAFO	North Atlantic Fisheries Organization
NASCO	North Atlantic Salmon Conservation Organization
NATO	North Atlantic Treaty Organization
NEAFC	North-East Atlantic Fisheries Commission
NFFO	National Federation of Fishermen's Organizations
NIFF	Northern Ireland Fishermen's Federation
NIFSA	North Irish Sea Fishermen's Association
OECD	Organisation for Economic Cooperation and Development
PESCA	Fisheries restructuring initiative
PME	Permis de mise en exploitation (licence to fish)
PO	Producer Organization (same as FPO)
QMV	Qualified Majority Voting
RIVO	The Netherlands Institute of Fisheries Research
SBF	Save Britain's Fish
SFC	Sea Fisheries Committee
SFF	Scottish Fishermen's Federation
SFO	Scottish Fishermen's Organization
SFPO	Shetland Fish Producers' Organization
SOAEFD	Scottish Office Agriculture, Environment and Fisheries Department
SPFA	Scottish Pelagic Fishermen's Association
SQM	Sectoral quota management
SWFPA	Scottish White Fish Producers Association

SWFPO	Southwestern Fish Producers Organization
TAC	Total Allowable Catch
TURF	Territorial Use Right in Fisheries
UAPF	Union des Armateurs à la Pêche de France
UNCLOS	United Nations Convention on the Law of the Sea

1 Fishing: a Defence of Politics[1]

Tim S. Gray

INTRODUCTION

This book originated in a colloquium on the politics of fishing that I organized on 17–18 September 1996 at Newcastle University. The aim of the colloquium was to bring together fishers,[2] officials, scientists and academics from several countries to discuss the issues which underlie the current crisis in the fishing industry, in the context of the charged political atmosphere within which decision-makers operate. The central (but not exclusive) focus of the colloquium was on the working of the European Union's (EU's) Common Fisheries Policy (CFP), which has become the subject of intense political debate, especially in the United Kingdom (UK), during the last three or four years.

The politicization of the fishing industry is nothing new. Fishing activity has always generated political controversy. The most extreme form of such controversy is, of course, a 'fishing war' and the UK and Iceland were locked into four so-called 'cod wars' in the 1950s and 1970s. This dispute was the first time that two liberal democratic states, both members of the North Atlantic Treaty Organization (NATO) had reached the stage of exchanging fire power over the issue of overfishing. The long-running crisis was precipitated by Iceland's successive declarations of a four-mile, 12-mile, 50-mile and 200-mile limit off its shores to protect its fishing industry which was experiencing a sharp fall in catches. The dispute was finally settled in December 1976 with the triumph of Iceland and the end of Britain's deep-sea fishing industry.

Although there have been no fishing conflicts amongst EU member states on the scale of the cod wars, there have been some quite serious local skirmishes over tuna fishing, for example between Spanish vessels on the one hand, and British and French vessels on the other. Moreover, there have been some serious political tensions between member states over fishing issues. For instance, Britain took Canada's side in 'its Greenland halibut war' with Spain in 1995. The question arises, therefore, why do such disputes occur? One reason is the open access nature of fish stocks, transcending national boundaries and thereby leading to clashes between fishers from different countries. Such clashes are whipped up by the media into international incidents (Leigh, 1983: 88). As Wise writes, 'fishery disputes can be easily, if erroneously, reduced to the simplistic, emotive issue of a national struggle against "foreign invaders" of "our" sea space'. (Wise, 1984: 1).

Another reason why fisheries disputes arise is because of the personality of fishers. As Cann points out in Chapter 2, fishers are rugged individualists for whom hunting is a way of life, and they are easily aroused in defence of their livelihoods. Moreover, fishing activity is often concentrated in peripheral regions which are both economically deprived and politically alienated (Leigh, 1983: 9), thus raising the temperature of intolerance.

But the main reason for conflict is the heightened competition for the increasingly scarce fish stocks. Scarcity of fish stocks is a world-wide problem. As the House of Lords Select Committee on Science and Technology observed in 1996: 'The world's fish stocks are in a state of crisis.' The Committee quoted the United Nation's Food and Agriculture Organization (FAO) which stated in its 1995 report on the State of the World Fisheries and Aquaculture that 'At the beginning of the 1990s, about 69 per cent of the world's conventional species were fully exploited, overexploited, depleted or in the process of rebuilding as a result of depletion' (House of Lords, 1996: 5). In some fisheries there have been spectacular collapses of stock. For example, the cod stocks in the previously abundant fishery off the Canadian Grand Banks were almost completely wiped out in the late 1980s, while the herring fishery of the North Sea had to be closed between 1977 and 1983, and has still not fully recovered. According to the Basis Report on Fisheries prepared for the Intermediate Ministerial Meeting on the Integration of Fisheries and Environmental Issues in 1997 (IMM 97), several species in addition to herring are currently considered to be 'outside safe biological limits' in areas of the North Sea – including North Sea plaice, cod and hake (CONSSO, 1997: 146–7).

There is a widespread perception that fisheries management has failed to address this problem of resource scarcity. Nowhere is this perception more acute than in relation to the CFP. A typical verdict is that of Karagiannakos: 'the current CFP is widely regarded as being a failure, mostly for allowing a decline in fish stocks and for being unable to deal with excess capacity in fishing vessels' (Karagiannakos, 1995: 1). But there is much less agreement on the reasons for this failure. In the chapters below we shall see many different diagnoses given for the failure of the CFP along with several alternative prescriptions for reforming or replacing the CFP. In the remainder of this introductory chapter, I analyse six perspectives which lie behind these various diagnoses and prescriptions: epistemological, psychological, instrumental, bureaucratic, corporatist and political. In the course of this analysis, I defend politics from the charge that it is mainly to blame for the failure of fisheries regimes.

In order to make more sense of these six perspectives, however, it may be helpful to draw on some of the ideas that have arisen out of two theoretical debates. The first debate lies in international regime theory, and is exemplified in Krasner's (1993) distinction between the liberal view and the realist view of regimes. On the liberal view, cooperation between states is a means of enabling each state to gain from mutual restraint in a shared endeavour. On the realist view, international regimes are zero-sum games or power struggles and are established and used to promote the interests of particular states.

The second debate lies in environmental theory, and is expressed in the literature on common pool resources (CPRs). Hardin (1968) famously originated the doctrine of the 'tragedy of the commons' by which he meant that the inevitable result of all users pursuing their own individual rational self-interest was the degradation and eventual destruction of the commons for all. Hardin's seminal article provoked a widespread debate on how to avoid this tragedy. Some writers (Heilbroner, 1974; Ophuls, 1977; Holden, 1994) argued that only a top-down or strong central authority – either national or international – could enforce the necessary cooperation upon users. At the other extreme, writers such as Demsetz (1967) and Smith (1981) claimed that privatization was the only answer – dividing the commons into individual lots that are sold or parcelled out to users. In between these two extremes there lies the more recently fashionable view held by Berkes (1989), Ostrom (1990), Bromley et al. (1992) and Keohane and Ostrom (1994) that local common users should be given the opportunity to establish their own collective/communal bottom-up solutions to avert the tragedy (as they have done in many parts of the world (McGoodwin, 1990: 92)).

As we shall see, many of these views are represented in the chapters that follow. For example, the liberal interpretation of regime theory is implied in Steel's account of the role of the European Parliament; the realist interpretation of regime theory informs Cann's reflections on the fisheries manager's perspective; the tragedy of the commons theory applies to much of Oliver's analysis of the quota system; the top-down solution of a strong central authority is critiqued in the paper by M. Gray; the privatization proposal is analysed in the O'Connor and McNamara chapter; and the devolution or bottom-up recommendation is the guiding light of several contributions, including those of Phillipson, Goodlad, Deas, Symes and Crean.

EPISTEMOLOGICAL PERSPECTIVE

The problem of managing fisheries under conditions of resource scarcity is the central theme that runs throughout all chapters in this book. The ways in which that problem is conceived can be categorized into six perspectives. The first perspective has an epistemological root. Here the scapegoat is the marine scientist who is accused by some fishers of working with a flawed methodology in assessing fish stocks. Some fishers claim from their own working knowledge that many species deemed by scientists to be declining are in fact increasing (Symes, 1996: 4), and they highlight the fact that the scientists themselves freely acknowledge the methodological hazards involved in their analytical predictions of fish stocks. As Professor John Shepherd (ex-Deputy Director of Fisheries Research, Ministry of Agriculture, Fisheries and Food (MAFF) at Lowestoft) put it, 'predicting fish stocks is like predicting the weather, only worse' (*Financial Times*, 24 November 1993). A margin of error of up to 50 per cent has sometimes been admitted. Even the head of the Advisory Committee on Fisheries Management (ACFM), the body responsible for giving the scientific advice to the EU Commission on which its proposals for the annual total allowable catches (TACs) are based, Jean-Jacques Maguire, said that scientific predictions of fish stocks are generally accurate only to within margins of plus or minus 25 per cent (*Fishing News*, 14 February 1997: 5).

Significantly, as Symes points out, during the last ten to twelve years scientists themselves have begun to question the Newtonian assumption on which the scientific theory of fisheries management has been based since the mid-1950s. This Newtonian assumption is that 'most fish stocks are inherently stable, behave predictably under moderate levels of exploitation and tend towards an equilibrium state'. Recent research has shown, however, that there is a 'tendency towards instability'. And Symes notes that 'chaos theory comes much closer to the fisherman's own experiential perceptions of Nature as unpredictable' (Symes, 1996: 5, 7). On this view, if there has been a collapse in a fishery, it would probably have been due to environmental circumstances which are beyond human capacity to understand, and which may very well go into reverse, restoring the fishery in due course: 'fish are fecund and have millions of eggs, which allows populations to "bounce back" rapidly.' (Middleton and Ellis, 1997: 12). The epistemological elements in this conflict of opinion between fishers and scientists are explored by Pálsson in Chapter 14.

PSYCHOLOGICAL PERSPECTIVE

The second perspective has a psychological root. Here the scapegoat is the greedy fisher. This is the classic tragedy of the commons situation, where fishers seek to maximize their own short-term interest and the result is that many stocks have become commercially extinct. As Shepherd puts it:

> There is a fundamental difficulty here that what is in the fisherman's long-term interests may not be in his short-term interests and the essential problem is that for an individual fisherman, his short-term interests are very often best served by evading or abusing the regulations and that is in the long-term interests of no one, but, human nature being what it is, it is understandable that a number of fishermen behave in ways which in the long term and collectively are seen to be irrational.
> (House of Lords, 1995b: 217; cf. Holden, 1992: 8)

Symes, similarly, claims fishers cling to the illusion propagated by T. H. Huxley a hundred years ago of the inexhaustibility of the sea's bounty. 'Despite all the evidence from more than a century of scientific investigation, the fishing industry appears to regard the resources not simply as renewable but also as inexhaustible. Fishermen still prefer to live by the discredited myth, in studied ignorance of the scientific facts' (Symes, 1996: 4). The psychology of fishers is discussed by Cann (Chapter 2), Oliver (Chapter 5) and Mikalsen (Chapter 7).

INSTRUMENTAL PERSPECTIVE

The third perspective has an instrumental root. Here the scapegoat is the set of tools used by management to regulate fishing. Particular criticism has focused on five sets of tools: the quota system, 'technical measures', decommissioning, days-at-sea legislation, and enforcement procedures. Let us examine these briefly in turn. On the quota system, as Oliver shows at length in Chapter 5, quotas encourage fishing up to the limit of the quotas; they lead to high rates of 'discards' (dumping of over-quota fish[3]); they tempt fishers to sell over-quota fish on the black market ('black fish'); and they raise the ethically intractable problem of how to allocate quotas between member states (a problem discussed at length by T. S. Gray in Chapter 15). Moreover, quotas undermine the individuality and self-esteem of fishers; indeed, they help to criminalize them, as Vestergaard points out:

> Quotas prevent the fishermen from freely using their usual strategies of action to cope with fluctuating stocks and prices. And this also prevents them from confirming their identity as valued examples of self-reliant members of society and as valued servants of the national household. In fact, the strategies of action that used to earn them recognition now criminalize them.
>
> (Vestergaard, 1996: 88)

In the case of the CFP, the quota system has also given rise to one of the most contentious issues currently exercising three of the member states in the EU – the issue of 'quota hopping'. The term 'quota hopping' or 'flagships' refers to the practice of nationals of one member state buying vessels of another member state and registering them in that other state. The purpose of doing this is to obtain a share of the quota allocated to the other country. But in British eyes, quota hoppers (which in tonnage comprise about a quarter of the UK fishing fleet) undermine the CFP's principle of relative stability which was established in 1983 to protect the national fishing interests of individual member states. In an attempt to stop quota hopping, the British government passed the Merchant Shipping Act (MSA) in 1988, but in the so-called

'Factortame' judgment in 1991, the European Court of Justice (ECJ) ruled against the British government on grounds that the Act violated the Treaty of Rome's principle of non-discrimination. (This was the first time that the ECJ had overruled a member state's Act of Parliament.) The British government subsequently promised fishermen that it would seek an EU ban on quota hopping at the Intergovernmental Conference (IGC) commencing in Turin in April 1996. Indeed, it vowed that it would not attempt to meet its targets for the Fourth Multi-Annual Guidance Programme (MAGP) until that ban was imposed. But the likelihood of an EU ban on quota hopping is very remote, and the British government's promise was seen more as a political ploy to deflect the anger of fishers from itself to the EU than as a serious policy objective.

Turning now to the so-called 'technical measures' used by management, criticism has focused on minimum mesh sizes (MMSs) and minimum landing sizes (MLSs). MMSs are very difficult to operate in a multi-species fishery, and unless an army of inspectors is employed to police every wholesale fish market, MLSs are impossible to enforce. Widespread dumping and illegal landing of juvenile fish is the result.

Another much criticized management tool is decommissioning. Decommissioning is compensation paid to fishermen for withdrawing their vessels from fishing. But critics have pointed out that its effect on fishing is small because the compensation terms offered are generally too low to be attractive to fishermen; that only the least efficient boats are offered for decommissioning; and that decommissioning merely increases intensification of fishing effort by the remaining vessels.

An alternative (or complementary) means of reducing fishing effort is by restricting the days at sea that vessels can fish. Some days-at-sea schemes do operate in the EU on a restricted basis, but the one introduced by the UK in 1992 had to be withdrawn because of furious objections from the industry. This scheme, which has been described as 'decommissioning on the cheap', provoked the fiercest opposition by UK fishers to any governmental measure ever introduced in Britain. It was described as 'the most controversial piece of UK fisheries legislation yet to have passed through the House of Commons' by Richard Banks, Chief Executive of the National Federation of Fishermen's Organizations (NFFO) (Banks, 1993: 15). The fishing industry was united in its resistance to the days-at-sea scheme, because foreign vessels would be out fishing while UK vessels were tied up.

In the face of such opposition, which led to a legal challenge by the NFFO in July 1993, referred by the High Court to the ECJ, the government withdrew the scheme in 1995 (even though the ECJ voted in its favour). The whole episode was a public relations disaster for the government. The days-at-sea legislation was a hastily concocted measure that was produced to fill an empty place in the parliamentary timetable shortly after the 1992 general election. It was an ill thought-out piece of legislation that had not been properly discussed beforehand with the fishing industry. It had two lasting effects. First, as M. Gray shows in Chapter 8, the government recognized the necessity for much greater consultation with fishers, and MAFF put in train a series of meetings to try to restore the confidence of the industry. Second, there has been an increased radicalization of fisheries opinion. It is no coincidence that the Save Britain's Fish (SBF) campaign, which was founded in 1990, rocketed into prominence at the time when the days-at-sea controversy was at its height. Extremist opinion had hardened and was not to be placated by the government's subsequently more softly, softly approach.

Finally, much criticism has been made of management failures to enforce fishing regulations. As implied above in relation to MLSs, the costs of effectively policing the fishing industry would be astronomical. With hundreds of fish markets in the EU, and tens of thousands of fishing vessels, the task of monitoring adherence to mesh sizes, landing sizes and quota regulations (including by-catches) is virtually impossible, without disproportionately high expenditure on equipment (including high-powered patrol boats), surveillance and labour. Given that, for example, in the UK the total value of fishing landings is only £550 million annually, there is a limit to the amount of enforcement expenditure the government is willing to commit beyond the present level of about £45 million.

BUREAUCRATIC PERSPECTIVE

The fourth perspective has a bureaucratic root. Here the scapegoat is the management elite that excludes fishers from participation in decision-making. Symes explains how over the last forty years, fishers have been gradually marginalized in the decision-making structures. As the forces of globalization, capitalization and centralization have begun to bite, traditional systems of management have given way to bureaucratic systems. 'Customary systems of management have been replaced by centralized bureaucratic policy making which, in most but not all instances, excludes the fishermen's organizations from an active influential role and has singularly failed to carry conviction and win support among the resource users.' As a result of this and other factors, says Symes, fishers have lost confidence not only in the system but also in themselves. And he sees a direct link between this loss of confidence and the disaffection which has led to direct action: 'The near anomic condition of the fishing community and its alienation from the management system has sparked off direct action and civil disobedience both on land and at sea ... the classic symptoms of institutional failure' (Symes, 1996: 9).

Two long-term consequences have followed from this marginalization process: first, fishers feel little moral compunction in breaking fishing regulations. Because these regulations were imposed without their consent or even consultation, they have no sense of commitment to them. Indeed, they justify their criminal behaviour on grounds that they are driven to it by the rules themselves; if they scrupulously observed the regulations, they would be bankrupt. Second, since they are excluded from the management decision-making process, they are left with no alternative but to exert external pressure on government ministers: 'the fishing industry is totally excluded from the formulation of management advice. The result ... is that the industry lobbies at the only level at which it can influence decisions, that of ministers' (Holden, 1994: 220).

As shown by M. Gray in Chapter 8 and by Phillipson in Chapter 9, until recently in the UK the government distanced itself from the fishers' organizations, restricting them at best to a purely consultative role. However, during the two years up to the 1997 general election, the extent of the contact between government and fishers in Britain greatly increased – even the Conservative Prime Minister met federation leaders. Another initiative taken by the Conservative government was to call for regional consultative committees. And it claimed the credit for some success at the EU level, in that as a result of a proposal from Britain, the European Commission arranged a series of meetings to take place in 1997 between fishers' representatives, scientists and administrators on a range of issues relating to North Sea fishing. Moreover, the EU has established an Advisory Committee on Fisheries (ACF) to provide for consultation with the fishing industry; it consists of representatives from all fishing industry sectors together with consumers.

These moves by the Conservative government and the EU suggest that managers, as well as fishers, are disillusioned with the existing structure of decision-making in the industry. However, we should not overestimate the momentum towards change in the structure of decision-making in the fishing industry. As M. Gray explains, the UK government's moves have been widely interpreted as stalling devices, designed to take the heat out of an increasingly fractious political atmosphere in the run up to the general election of 1997, rather than a genuine attempt to take the industry into a decision-making role. And, according to Holden, the EU's ACF has been of only limited help in linking fishers to managers: it meets infrequently; its subcommittees have turned out to be mere talking shops for the rehearsal of national interests; and members of the subcommittee on resources are angry that they are often not consulted on proposals until the Commission has accepted them. (Holden, 1994: 211–12). Also, as Sandberg (1996: 36) points out, most managers are still wedded to the existing structure.

Moreover, there are two obstacles to participation on the fishers' side. First, fishers are fragmented, both geographically and sectorally, and it is not easy to see how to organize their participation in decision-making. For example, in the UK, there are two federations (NFFO for England and Wales and the Scottish Fishermen's Federation (SFF) for Scotland), 15 producers' organizations and over 60 local associations, each with its own particular interests (Holden, 1994: 218). Second, fishers' representatives are unhappy about becoming too closely involved with government processes of decision-making, since they risk being blamed by both government and their fellow fishers if things go wrong.

Nevertheless, it is frequently argued that until and unless fishers are given more responsibility for the decisions that govern their lives, the wrong decisions will continue to be made and fishers will continue to regard the regulations as alien rules, to be evaded not respected. In several of the chapters which follow, alternative decision-making structures are considered, including sectoral quota management by fish producer organizations (FPOs) (Chapter 10), coastal state management (Chapter 13), regionalization (Chapter 12) and sea territories (Chapter 11).

CORPORATIST PERSPECTIVE

The fifth perspective has a corporatist root. Here the scapegoat is the weak fisheries minister, unable to withstand pressure from the fishing industry to resist TAC reductions. As Lord Selborne noted at a House of Lords Science and Technology Committee meeting, 'It seems that the advice given by scientific experts over the years on sustainable fishing yields has, almost invariably, not been adopted by the Fisheries Ministers.' Alain Laurec, Director of Conservation and Environmental policy in DG XIV (the Fisheries Directorate) agreed that this statement was 'largely true', and that it represented not only the tragedy of the CFP, but 'the tragedy of all fishery policies' (House of Lords, 1995a: 2). Ad Corten, an ex-member of the International Council for the Exploration of the Sea's (ICES) Advisory Committee on Fisheries Management (ACFM) has pointed out that fisheries ministers are not professionally qualified and they have short political horizons. 'Their main objective is to avoid short-term political problems, and to show fishermen that they are trying to help them. What happens in the long run is probably somebody else's problem' (Corten, 1996: 10). Significantly, UK fisheries ministers have a very brief shelf life: there were three in the last four years of the Conservative government – David Curry, Michael Jack and Tony Baldry. Thom and Phillipson touch on the role of British, French and Danish fishery ministers in Chapters 4 and 9.

POLITICAL PERSPECTIVE

The sixth and last perspective has a political root. Here the scapegoat is the politician who treats fishing issues as pawns by which to advance wider political objectives which have nothing to do with fishing. On this view, which often lies behind the previous five perspectives, policy decisions about fisheries are taken essentially for political rather than conservation reasons. Indeed, in some instances, fisheries issues are merely political footballs, kicked around by states to gain some leverage in other policy areas. According to this perspective, politics has hijacked the fisheries agenda, and the issue that should be centre stage – conservation of fish stocks – has been peripheralized. As Austin Mitchell (Labour MP for Great Grimsby) put it, 'Fishing is far too important to be left to the politicians or to be sacrificed to the EU' (*Fishing News*, 25 November 1994: 5).

The most forceful advocate of this view is Holden, who gives many illustrations of the dominance of politics over conservation in decision-making in the CFP. For example, the system of TACs was not introduced to protect fish from a dangerous level of exploitation, but to provide a distributive mechanism for sharing out fish stocks between member states. 'The primary objective in establishing a system of TACs under the CFP was political; the member states wanted to share the fish resources amongst themselves and the only means by which they could do this was by a system of TACs allocated by national quotas' (Holden, 1992: 13). According to Holden, fisheries decisions are reached by a process of political 'horse trading', not a process of rational management of stocks. This is why CFP legislation is 'complex, often contradictory, difficult to implement and impossible to enforce' – because it is forged in a process of political brinkmanship (Holden, 1994: 214).

Paradoxically, a similar position is often adopted by fishers, arguing that the intrusion of political considerations has destroyed their livelihoods. The charge here is that the fishing industry, or at any rate some segment of it, is being undermined by the pursuit of non-fishing political objectives. For British fishers the most important illustration of this charge was the entry of Britain into the EC in 1973. The critics, such as the Save Britain's Fish (SBF) campaigners, who want Britain to leave the CFP, argue that Prime Minister Edward Heath, in signing the Treaty of Accession in January 1972 which committed Britain to join the EC, surrendered the bulk of the waters around the British Isles to other member states. This surrender was one of the prices Britain had to pay in order to persuade France to lift her veto on British membership and to obtain access to EC markets for commonwealth food exports. As Leigh explains:

> The British government was not so enamoured of the Community that it would accept any entry terms. But it refused to permit a minor issue such as fisheries to hold up signature of the Treaty of Accession. Fisheries were far less important in Britain than in Norway. Without the prospect of a referendum providing a forum for the dissemination of views hostile to the Community, the British government felt under less pressure than the Norwegian government to obtain terms of entry which would be acceptable to all sectors of the fishing industry.
>
> (Leigh, 1983: 41)

Similarly, the Wilson government played down the fisheries issue in the referendum in 1975 on the renegotiated terms of British entry to the EC (Boardman, 1976: 194–5).

There are many other examples of the sacrifice of fishing interests to serve broader political objectives. For instance, in 1994 Spain successfully threatened to veto the Enlargement Accord to ratify the entry of Austria, Finland and Sweden into the EU unless it was granted greater access to 'European waters' seven years earlier than previously agreed. The CFP's sympathetic treatment of Spain, not only by cutting short the transitional period before full access to EU waters, but also by providing generous grants for rebuilding and modernizing its fishing fleets and by turning a blind eye to its poor enforcement of MLS rules, is seen by British critics as part of a wider political agenda to lock Spain into the EU as a secure liberal democratic state. Given the very high rates of unemployment in Spain and the strong sense of regional identity in fishing areas such as the Basque country, the health of the Spanish fishing industry is regarded as a crucial factor in the continued stability of Spain's democratic political system.

Another EU political objective that British Eurosceptic critics see fishing interests being sacrificed for is the creation of an EU super state. Enoch Powell raised this spectre in 1975 (Boardman, 1976: 194) and twenty years later Booker and North rehearse the same theme, claiming to have discovered DG XIV's 'Ground Plan' or secret strategy to achieve an 'EU Fisheries Policy' to replace the CFP, in which there would be no national quotas and no principle of relative stability. Booker and North characterize this objective as 'another EU Grand Design, the fisheries equivalent of the Single Currency ... the plan ... to merge all the fishing fleets of Europe by the year 2003 into one "Union fleet" fishing "European Union waters" under the central control of Brussels' (Booker and North, 1996: 83, 193–4). Similarly, John Ashworth, SBF's conservation spokesperson, characterizes it as the advance of federalism: 'What we are facing is, fishing is becoming a much bigger issue than purely fishing. What it is all about is where Europe is going, that is why it is such a political thing now. It's all about, do you agree with a federal Europe or do you agree with the independence of states?' (Ashworth, 1996: 4).

Such sentiments mark a radical departure from the traditional forms of political engagement by UK fishers' organizations. This shift can be characterized as a move away from interest group politics to pressure group politics. The main representative organizations – SFF in Scotland and NFFO in England and Wales – serve essentially as professional interest groups, seeking to persuade governments to recognize the particular problems of the industry and to take the appropriate steps to address them, especially by resisting downward pressure on TACs at Fisheries Council meetings. But the SBF has stepped up the issue into a broader campaign against the CFP as a whole, and it has forged links with political allies such as Eurosceptic MPs to attract a wider public audience.

Capitalizing on the widespread discontent felt by British fishers during the days-at-sea conflict, the SBF rapidly gained support. Its objectives were endorsed by several fishers' organizations, including the Scottish White Fish Producers Association (SWFPA), the Southwestern Fish Producers' Organization (SWFPO) and (in a qualified form) the NFFO. In 1995 a number of senior figures in the SWFPA (which had cooled in its enthusiasm for withdrawal from the CFP) left to form a new organization, the Fishermen's Association Limited (FAL), with a central platform of support for the SBF campaign. In a statement in *Fishing News* (25 October 1996) the FAL declared that 'Save Britain's Fish, which FAL supports, has put fishing on the political agenda.... Fishing before had no political clout.'

However, the SBF has not attracted the support of Britain's most powerful federation, the SFF. On the contrary, the SFF has been the SBF's most bitter opponent, accusing it of pursuing a will-o'-the-wisp strategy, since there is no conceivable prospect of Britain withdrawing from the CFP. Neither of the two main UK political parties is willing to endorse withdrawal; both of them want to reform, not leave, the CFP.[4] In a submission to the Parliamentary Maritime Group Advisory Meeting on 'The Future of the EU Common Fisheries Policy', Bob Allan (Chief Executive of the SFF) wrote that 'we must seek changes from within the system rather than to support a fruitless campaign to withdraw from it' (Allan, 1996).

But the SBF is undaunted by such criticisms, pointing out that political configurations are not unchangeable. As Ashworth said in 1993, 'if political pressure is applied ... who knows what's possible? The situation has changed so dramatically in the last six months. Who would have said we could come out of the ERM? They denied that right up until the last moment. So what cabinet ministers say today cannot really be believed' (Ashworth, 1993). The SBF strategy is long term, to raise public awareness of the issues and to wait for a groundswell of support that will compel politicians to act. How effective has been the SBF campaign? It is difficult to say. On the one hand, the SBF has been accused of dividing the industry into sharply polarized and mutually recriminatory camps – the radicals, who want withdrawal from the CFP, and the reformists who want change in the CFP – therefore sending a mixed message to London and Brussels. On the other hand, the SBF has certainly raised the political profile of the fishing industry in the UK, and in the pre-election atmosphere before May 1997 this helped to extract some concessions from the Conservative government (such as more funds for decommissioning). It may also have stiffened the British government's stance on quota hopping at the IGC.

So far in this analysis of the sixth perspective on the fishing crisis – that the crisis is due to the way in which politics has intruded into fishing – most of the points have been derogatory to politics. The underlying assumption has been that the intrusion of politics into fishing has resulted in damaging decisions being made in relation to the health of fish stocks. But we must now turn to a more positive picture. For one thing, there are writers who argue that politics has had a benign effect on at least some fish stocks. For example, even such a trenchant critic of the CFP as Holden affirms that in respect of three species – herring, mackerel and plaice – the CFP 'has been ostensibly very successful' (Holden, 1994: 156).

Moreover, several writers point to the success of the CFP in its relationship with third countries. Leigh claims that success in this area is due to the fact that it is easier to get agreement on external policies than on internal policies.

> The external relations of the European Community are partially immune to the conflicts which often block progress on internal issues. It is easier to unite in adopting a common position towards third countries than it is to agree on common policies which distribute resources among member states. Political leaders operate in a relatively permissive environment when dealing with issues beyond the horizon.   (Leigh, 1983: 119)

For another thing, it can be argued that without the CFP, European fish stocks would be even more depleted than they are now. The existence of the CFP has imposed at least some discipline upon the exploitation of fish stocks by member states. Moreover, the CFP, which has survived in its present form since 1983, has held fast to the principle of relative stability and the safeguarding of coastal states' interests (through the 6- and 12-mile limits, the Hague Preferences and the Shetland and Irish Boxes).

Finally, it could be claimed that the wider political objectives served by the CFP justify the sacrifice of at least some fish conservation objectives. Symes and Crean contrast these two alternative sets of objectives as follows:

> The underlying principles of the CFP – horizontality, non-discrimination and relative stability – have more to do with reinforcing the concept of European unity and co-operation than with effective management of a seriously depleted, highly sensitive and unstable resource. The CFP is a political statement, neatly aligned with the Community's general principles and designed to avoid rocking the European boat. It seeks, therefore, to reinforce economic and political stability within the Community – a precept which translates uneasily into a policy framework and regulatory system. As a result, the ensuing system is singularly ill-suited to the particular conditions of Europe's fisheries – mixed fisheries within which the component species exhibit different biological structures and characteristics and respond differently to specific regulatory measures.   (Symes and Crean, 1995: 398)

While Symes and Crean do not indicate whether they believe it is a price worth paying, it could be argued that European unity and cooperation is more valuable than fish conservation. However, my own conclusion is that whether or not European unity is judged to be more valuable than fish conservation is itself a political issue. If fishers or scientists or managers or environmentalists or academic writers deplore the influence of politics on fisheries decisions, they should reflect that in the last analysis fisheries decisions are necessarily political decisions. For example, it is not accurate to say that if TACs are raised, it is for political reasons, whereas if TACs are lowered it is for conservation (that is, non-political) reasons. Both decisions are political: in the first case, priority is given to non-conservation objectives; in the second case, priority is given to conservation objectives. But in each case the prioritization is a *political* choice.

It is no use objecting, therefore, to 'political' considerations intruding: political considerations can never 'intrude' since they can never be absent in any decision-taking process. Those who object to political considerations intruding are simply expressing their disappointment that in the necessarily political debate, their view did not prevail. In this respect, fisheries policy is like any other sectoral policy – bound up with politics. The proper response of those who feel that inadequate weight is being paid to conservation considerations is not, therefore, to cry 'foul' because politics is illegitimately intruding, but to campaign more vigorously for their point of view. This is precisely what growing numbers of environmentalists are doing – engaging explicitly and openly in the political debate to persuade the public (especially consumers of fish products, both household and industrial) to put economic pressure on fisheries ministers to adopt a more conservation-minded set of regulations designed to protect fish stocks. It is also what the SBF and the FAL are doing, pursuing in public debate the case for 'repatriation' of Britain's fish stocks. In both cases the objective is to ensure that the political costs of, respectively, not conserving fish or not repatriating fish stocks are too great for fisheries ministers to ignore. Whatever we may think of either of these particular objectives, we should acknowledge that the willingness of environmentalists and the SBF/FAL to engage in political activity is a healthy, not an unhealthy sign in an open, pluralist, democratic society. The North Sea Conferences exemplify this position: 'The North Sea Conferences are political events . . . [they] have provided the vehicle for political cooperation' (Andersen and Niilonen, 1995: 32). And as an editorial in *Fishing News* pointed out, 'political institutions . . . (are) what fishermen's trade organizations essentially are' (*Fishing News*, 8 September 1995: 2).

On this view, the implicit desire of writers such as Holden to take politics out of fisheries management is fundamentally misconceived. Holden argues that 'the majority of Member States do not take their responsibilities seriously and, in consequence, the whole policy is failing. For management to be effective, responsibility cannot be divided. Management cannot be left to the Member States. Management must be the responsibility of the Commission; that is its institutional role' (Holden, 1994: 245). The implicit assumption behind Holden's view is that fisheries management is a matter of top-down administration (a technical matter best left to scientific experts) not a matter of bottom-up politics (whereby ultimately it is public opinion that determines the priorities). Such a view is not only undemocratic but it ignores the fact that scientific elites are themselves political actors, and that their views are shaped just as much by political value judgments as are the views of the rest of the population. As Stairs and Taylor observe of the London Dumping

Convention (LDC), 'Unfortunately, few of the government-appointed scientists, in our experience, have been aware that value-free science was long ago discredited, and they still have a self-image of "scientific judgements", which they make, and "political" decisions, which others make (be they parliaments, the consultative committee of the Convention or even environmental groups)' (Stairs and Taylor, 1992: 118). The truth is that fisheries issues are necessarily and inescapably political issues; fishing regimes are ineluctably political organizations; and the policies they pursue are inevitably a reflection of political bargaining processes. We cannot take politics out of fishing policy any more than we can take politics out of agricultural policy or transport policy or health policy or education policy or any other policy area, nor, in my view, should we wish to do so. As Sandberg puts it:

> Fishing activity has always been dependent on politics. . . . It can be argued that politics has shaped the entire pattern of fishing hamlets along the coasts of the Atlantic Ocean, the Mediterranean Sea and the Baltic Sea, out of a concern for the relations between fisheries and other agents in society. . . . What we today call traditional fisheries have been shaped by the political history of Europe.   (Sandberg, 1996: 35)

OUTLINE OF THE CHAPTERS

Chapter 2 is an expression of the fishery manager's perspective from Charles Cann, who stresses the concept of the 'tragedy of the commons' and the resulting 'process of attrition' in the 'battle between fishermen and managers'. Noting the unique psychology of fishers and the clash of culture between them and managers, Cann refers to 'a lifetime of crisis management'. The EU adds another layer of complication, wherein responsibility for decisions is often fudged, not least because member states alone have power to enforce the regulations. Cann concludes on a pessimistic note by warning that there is no simple panacea for our fisheries problems, and that the necessary steps to restrain fishing effort may not be taken until and unless there is a deepening crisis of fish stocks.

In Chapter 3 David Steel examines the often overlooked but increasingly important role of the European Parliament (EP). He points out that only in the EP are there systematic public debates about European fisheries policy. The potential significance of these debates is growing, both because fishers' organizations are increasingly targeting MEPs, and because the Commission and the Council are giving increasing attention to the EP's views, especially those expressed in the recently established Fisheries Committee.

In Chapter 4 Mireille Thom turns the spotlight on the fishing policies of two member states, in a comparison of Britain and France. Her findings include the unexpected conclusion that despite the very different approaches to industry in the two countries – Britain's laissez-faire approach contrasted with France's interventionist approach – their fisheries policies are broadly similar, both in style (authoritarian) and in substance (inadequate implementation of CFP regulations; half-hearted decommissioning schemes). Although there are some policy differences – for example, market forces in Britain largely determine licence and quota allocations, whereas in France licences and quotas are allocated by the state – for the most part, 'policies and outcomes are determined by sectoral imperatives'.

Focusing attention on one of these policy measures, Tim Oliver in Chapter 5 offers a highly critical analysis of the TAC/quota system which is used by EU member states for allocating shares of fish stocks. Oliver argues that the TAC/quota system was designed for distributive, not conservation, reasons, and it has served to exacerbate, not relieve, pressure on fish stocks. Discards, 'black fish', 'paper fish' (or 'ghost fishing'), catch misreporting and criminalization of fishers all result directly from the TAC/quota system.

One possible solution to these problems is a system of private property rights in individual transferable quotas (ITQs). In Chapter 6 Robert O'Connor and Brendan McNamara rehearse the case for ITQs, and examine the working of ITQ systems in New Zealand, Iceland, British Columbia and the Canadian Bay of Fundy. They conclude that while the ITQ system does not provide a general panacea for fisheries problems, it has worked well in several fisheries, and they recommend its introduction in the EU, starting with pelagic fisheries, though with strict regulation to prevent very high entry costs, the concentration of quotas into few hands, quota busting, high grading and by-catch dumping.

Management structures occupy the attention of the next seven chapters. Knut Mikalsen in Chapter 7 addresses the central issue of how to ensure grass-roots approval for painful decisions in the context of Norwegian fisheries management. Adequate representation is often held to be the sine qua non of legitimacy and compliance. However, Mikalsen suggests that the establishment of an independent and neutral body, removed from the political arena, with no representation of user groups, could deliver policies which fishers would accept. On the other hand, in Chapter 8 Mark Gray discusses recent moves by the UK government to get fishers more involved in decision-making, and asks whether this indicates a genuine desire for a more centralized style of management or a cynical attempt to win fishers' votes in the run-up to the 1997 general election. In Chapter 9 Jeremy Phillipson compares the fishing policy systems of Denmark and the UK, noting that, while both are centralized, the Danish system allows for greater involvement of user groups in policy-making. Indeed, Danish fishers' organizations are generally integrated into the policy formulation process: a 'participant' culture exists. By contrast, in the UK, there is only a 'subject' political culture. However, Phillipson points out that at the implementation stage, the UK fishing industry is much more involved than is its Danish counterpart.

John Goodlad in Chapter 10 examines one important example of such involvement at the implementation stage – the Shetland Fish Producers Organization's system of sectoral quota management (SQM). The SQM system allows fishers, through their producer organizations, to manage fisheries. Goodlad discusses the advantages and disadvantages of the SQM system, and concludes that it could develop into either a system of ITQs, or one of communally owned quota. In Chapter 11 Kevin Crean's advocacy of 'Sea Territories' bears some similarity to SQMs, in that both entail devolving responsibility for managing fish stocks to local users. Crean takes his inspiration from the territorial use rights in fisheries (TURFS) in the Pacific Basin, which for two thousand years secured the sustainable exploitation of marine resources. Crean points to several elements of TURFS in the CFP (such as the 6–12-mile limits, the Mackerel Box of South West England, the Norway Pout Box and the Shetland Box). Further extensions of TURFS could be justified by the EU's principle of subsidiarity, and could entail regional organizations to coordinate local users from several member states.

David Symes takes up this regional theme in Chapter 12, arguing for a more flexible, less monolithic and standardized approach to fisheries management in Europe. Reform of the CFP should take the form of establishing 'a comprehensive series of Fisheries

Councils covering the EU's regional seas' which would have the power to make all management decisions within their regions, subject to the overall authority of the European Commission and the Council of Ministers.

Barrie Deas, however, in Chapter 13, argues for a more radical break with the CFP: instead of regionalizing its powers, he proposes the nationalizing or repatriating to member states of complete authority over their coastal waters. Deas's radical proposal reflects the NFFO's fear that the CFP is becoming increasingly centralized and that only by giving to member states authority over their 200-mile limits, under a system of interlocking coastal management regimes, will their fishing industries have 'some assurance of a viable future'.

The two final chapters address epistemological and ethical issues respectively. In Chapter 14 Gísli Pálsson, focusing on Iceland, 'examines changes in the relative importance of the ecological knowledge of fishers and professional marine biologists in public discourse'. He argues that there is good reason to bring the practical knowledge of fishers back into the decision-making process of fisheries management, as the modernist notion of the 'absolute objectivity of science' is losing its credibility.

In Chapter 15 Tim Gray discusses the normative dimensions of the CFP, examining issues of distributive justice raised by its twin principles of equal access and relative stability. He argues that, while there is considerable doubt about the fairness of each of these principles, any verdict on the overall justice of the CFP must rest on complex considerations of both the categories of persons to whom we judge that the CFP should be fair, and the wider rules governing the relationship between member states in the EU.

NOTES

1. I am grateful to Anthony Stenson and Linda Cotterrell for conducting interviews in 1993 and 1996 respectively, which I have drawn on in this introductory chapter. I am also grateful to Ella Ritchie, Tony Zito and Mark Gray for their comments on an earlier version of this chapter.
2. The term 'fishers' is used instead of 'fishermen' since some fishers are female.
3. Holden points out, however, that discarding did not originate with the TACs/quotas system (Holden, 1994: 199).

4. It is worth noting, however, that the 1997 Liberal Democrat manifesto proposed 'scrapping the Common Fisheries Policy and replacing it with a new Europe-wide fisheries policy based on the regional management of fish stocks' (Liberal Democrats, 1997: 58).

BIBLIOGRAPHY

Allan, R. (1996) Contribution from the Scottish Fishermen's Federation to the Parliamentary Maritime Group Advisory Meeting on *The Future of the EU Common Fisheries Policy*, 13 February 1996.

Anderson, J. and Niilonen, T. (1995) *Progress Report 4th International Conference on the Protection of the North Sea, Esbjerg, Denmark, 8–9 June 1995* (Copenhagen: Ministry of the Environment and Energy, Danish Environmental Protection Agency).

Ashworth, J. (1993) Interview, 5 August 1993.

Ashworth, J. (1996) *The Save Britain's Fish Campaign* (unpublished transcript of paper delivered September 1996).

Banks, R. (1993) *Memorandum to the House of Commons Agriculture Committee* (Grimsby: NFFO).

Berkes, F. (1989) *Common Property Resources: Ecology and Community Based Sustainable Development* (London: Belhaven Press).

Boardman, R. (1976) 'Ocean Politics in Western Europe', in Johnston, D. M. (ed.), *Marine Policy and the Coastal Community* (London: Croom Helm).

Booker, C. and North, R. (1996) *The Castle of Lies: Why Britain Must Get Out of Europe* (London: Duckworth).

Bromley, D., Feeny, D. and Blomquist, W. (eds) (1992) *Making the Commons Work: Theory, Practice, and Policy* (San Francisco: Institute of Contemporary Studies Press).

CONSSO (1997) *Basis Report on Fisheries and Fisheries Related Species and Habitat Issues,* Prepared by the Committee of North Sea Senior Officials for the Intermediate Ministerial Meeting on the Integration of Fisheries and Environmental Issues in 1997 (IMM 97).

Corten, A. (1996) 'The Widening Gap between Fisheries Biology and Fisheries Management in the European Union', *Fisheries Research*, 27: 1–15.

Demsetz, H. (1967) 'Towards a Theory of Property Rights', *American Economic Review*, 62: 347–59.

Hardin, G. (1968) 'The Tragedy of the Commons', *Science*, 162: 1243–8.

Heilbroner, R. L. (1974) *An Enquiry into the Human Prospect* (New York: Norton).

Holden, M. (1992) *The Future of the Common Fisheries Policy* (Guildford: World Wide Fund for Nature).

Holden, M. (1994) *The Common Fisheries Policy Origin, Evaluation and Future* (Oxford: Fishing News Books).

House of Lords (1995a) Select Committee on Science and Technology (Session 1994–95), *Fish Stock Conservation and Management* (Evidence Received up to 11 July 1995), 10 May 1995 (London: HMSO).

House of Lords (1995b) Select Committee on Science and Technology (Session 1995–96), *Fish Stock Conservation and Management* (Evidence Received after 11 July 1995), 21 November 1995 (London: HMSO).

House of Lords (1996) Select Committee on Science and Technology (Session 1995–96), Second Report on *Fish Stock Conservation and Management*, 18 January 1996 (London: HMSO).

Karagiannakos, A. (1995) *Fisheries Management in the European Union* (Aldershot: Avebury).

Keohane, R. O. and Ostrom, E. (eds) (1994) *Local Commons and Global Interdependence* (London: Sage).

Krasner, S. D. (1993) 'Sovereignty, Regimes and Human Rights', in Rittberger, V. (ed) *Regime Theory and International Relations* (Oxford: Clarendon Press).

Leigh, M. (1983) *European Integration and the Common Fisheries Policy* (London: Croom Helm).

Liberal Democrats (1997) *Liberal Democrat Manifesto* (London: Liberal Democrats).

McGoodwin, J. R. (1990) *Crisis in the World's Fisheries* (Stanford, Calif.: Stanford University Press).

Middleton, C. and Ellis, A. (eds) (1997) *Report of Proceedings of UK IUCN Sustainable Fisheries Seminar*, 13 December 1996 (London: DoE).

Ophuls, W. (1977) *Ecology and the Politics of Scarcity* (San Francisco: Freeman).

Ostrom, E. (1990) *Governing the Commons* (Cambridge: Cambridge, University Press).

Sandberg, A. (1996) 'Community Fishing or Fishing Communities?', in Crean, K. and Symes, D. (eds) *Fisheries Management in Crisis* (Oxford: Fishing News Books).

Smith, R. J. (1981) 'Resolving the Tragedy of the Commons by Creating Private Property Rights', *Wildlife*, 1: 439–68.

Stairs, K. and Taylor, P. (1992) 'Non-Government Organizations and the Legal Protection of the Oceans: a Case Study', in Hurrell, A. and Kingsbury, B. (eds), *The International Politics of the Environment* (Oxford: Clarendon Press).

Symes, D. (1996) 'Fishing in Troubled Waters', in Crean, K. and Symes, D. (eds) *Fisheries Management in Crisis* (Oxford: Fishing News Books).

Symes, D. and Crean K. (1995) 'Historic Prejudice and Invisible Boundaries: Dilemmas for the Development of the Common Fisheries Policy', in Blake, G. H., Hildesly, W. J., Pratt, M. A., Ridley, R. J. and Schofield, C. H. (eds), *The Peaceful Management of Transboundary Resources* (London: Graham & Trotman) pp. 395–411.

Vestergaard, T. (1996) 'Social Adaptations to a Fluctuating Resource', in Crean, K. and Symes, D. (eds), *Fisheries Management in Crisis* (Oxford: Fishing News Books).

Wise, M. (1984) *The Common Fisheries Policy of the European Community* (London: Methuen).

2 The Politics of Fishing: a Fisheries Manager's Point of View

Charles Cann

INTRODUCTION

When I was invited to prepare this chapter I thought about en-titling it 'Why are fisheries managers so hopeless?' Having been involved directly and indirectly in fisheries management in MAFF for an uncomfortably large part of the last twenty years, I didn't feel like striking too upbeat a note about what might be my last semi-professional involvement with this bed of nails. However, in these politically highly charged times I thought that such a title might be misconstrued and taken literally and that would be unfair on my fellow fisheries managers who are still going to be strug-gling with these problems in the months and years ahead. In fact, in my view, the failures in fisheries management where they occur, and they do occur rather often, are in general not just down to fisheries managers. From one point of view, one could say that they are the collective responsibility of all concerned, including the fishing industries. But from another point of view, which I think I prefer, it seems to me that the difficulties are to a large extent the almost inevitable result of the conflicting forces at work in those fisheries where the management has not been a success. I some-times think that the surprising thing is where you find a fishery which is being managed successfully.

I shall begin by defining some of my terms. By 'fisheries man-ager' I mean to denote the politicians and administrators like my-self who take the decisions about how fisheries are regulated. I distinguish them from the scientists and other experts who provide a lot of the advice which underpins the managers' decisions. Of course for the UK fisheries, the managers include the Council of EU Fisheries Ministers and the Commission.

I start from the premise that the aim of fisheries management is

24

to provide a regulatory framework which delivers reasonably stable fisheries, subject to unavoidable biological variability, and enables them to be exploited in a reasonably efficient and profitable manner. In many fisheries around the world, including the fisheries around the UK, this aim has been far from achieved.

I want to offer some explanations of the main factors which I think account for the failures; and, in doing so, I want to distinguish between the difficulties which seem to be inherent in all fisheries management, those which are specific to the conduct of the EU CFP and the difficulties which impact particularly on the UK manager.

Starting at the general level, I would highlight two particular phenomena. First, there is the nature of the fishing activity, and second, there is what I would call the nature of fishermen. Both make the task of the fisheries manager peculiarly difficult.

THE NATURE OF THE FISHING ACTIVITY

By the nature of the fishing activity, I refer to its inherent tendency to anarchy or, in the more conventional phrase which we come across so often, the tragedy of the commons. It is an economic activity which exploits an economic resource in which, for the most part, no individual or individual enterprise has a readily established property right. In most fisheries where there are a large number of competing participants, most of them being pretty small, personal businesses, all participants tend rightly to believe that it is in their interest to exploit the resources as intensively as they can because they cannot count on benefiting from exercising restraint and husbanding the resource. By definition the fisheries manager is directly opposed to such anarchy which, if allowed its head, can destroy the industry and the resource or at least lead to major instability and inefficiency.

In so many cases, the resulting conflict between the fisherman and the manager is unusually stark. Managers and regulators of other economic activities can usually count on support from a worthwhile proportion of those participating in the economic activity, because quite often the manager's activities are seen as necessary and effective in providing fair terms of competition and protecting established property rights and so on. In the case of many, if not most, fisheries, managers do not have any overt support from any vested interest in the industry. The occasional support of environmentalists has, so far, done little to redress the balance although this could change if some recent developments take root and commercial organizations start to take more of an interest in the state of the fisheries from where they get their fish.

In the ensuing battle between fishermen and managers there is a process of attrition. For the fisheries manager to be able to impose his will in the difficult situations where fishing capacity and effort are excessive, the fisheries manager needs to be able to deploy massive resources on administration and enforcement, prosecutions and so on. In normal circumstances the political will and resources to see such a battle through to success will not be forthcoming. The political price in terms of public expenditure, fishermen's protests and lost jobs will rarely be worthwhile, bearing in mind that the good to be achieved – stock recovery – is neither very sexy nor very certain.

The fisheries manager either has to get in first, managing a newly established fishery, access to which is limited from the start, an access which he can rent out in a controlled manner; or he must wait for a major stock collapse that demonstrates that there is a crisis and a need for proper management and then it is possible that the necessary resources may be forthcoming to buy out excess capacity, develop alternative employment and so on.

To my mind, the calls for decentralized, regionalized or bottom-up fisheries management as a way out of this dilemma are something of a distraction. I can well see the attractions of defining local or regional fisheries and allocating management responsibility to those involved, internalizing, as it were, the fisherman vs manager conflict. But where you have seriously over-exploited fisheries as we have around our shores, I think you have got to solve the over-exploitation problem first.

THE NATURE OF FISHERMEN

The second general phenomenon I mentioned at the start was the nature of fishermen. This is not perhaps as important as the nature of the fishing activity nor perhaps quite as universal. However, I think it is quite an important factor in the problem of fisheries management that on the one hand you have typically the men in grey suits – like me – and the chaps in white coats – the fisheries scientists – and on the other the men in sea-going kit doing what is seen as real, rather romantic and dangerous work. The two groups do not naturally feel comfortable with each other and the chaps in the sea-going gear tend to get the better press and more popular support. A former UK fisheries minister had a rather unfortunate habit of saying that the one good thing about dealing with fishermen was that they made farmers seem so reasonable. There is a culture clash here which does not make fisheries management any easier, and that is not an insignificant point if you bear in mind that any management regime can succeed only if you have a shared view, among the managers and at least a decent proportion of the managed, that the regime is fair and reasonable.

In short, in fisheries management, the main participants, the fishermen and the managers, seem rather often to have very little in the way of common aims or common values, and that is a fairly tricky foundation. It is particularly unfortunate for an activity like fisheries management which is so fundamentally political.

Moreover, the nature of fisheries and fishermen is such that the managers tend not to get support from third parties such as the public and the media. If government tackles big business for some environmental shortcoming, the public and most of the media are instinctively ready to support the regulators. Fishermen are, however, generally perceived by the public and press sympathetically, and although nobody wants fish stock destroyed, the state of our generally unseen fish stocks has not yet prompted public concern in the way that, for example, whaling has done.

This creates a situation where all too often, the advantages for the politician-cum-manager of getting on top or even ahead of the management problems tend to be at least balanced by the controversy and odium which is likely to be involved in getting there. Moreover, this disincentive to decisive action is compounded by the uncertainties about fisheries. We say that fish stocks are under pressure, but we do not (and could not honestly) say that stock a, b and c will collapse by year x unless some specific action is taken. There is also the further and important uncertainty factor concerning the success of any new fisheries regime: given the ingenuity and contrariness of fishermen and the sheer complexity of the issues and activity involved, there is always in the heart of any fisheries manager the doubt about how well a new regime will work. There is the consequent fear that the upheaval of massive change will not deliver sufficient real gain to make the effort worthwhile.

All these factors conspire to inhibit substantially among managers and especially among democratically accountable politicians any enthusiasm for decisive, radical action. This is not a criticism of politicians. It is an objective assessment of the line-up of political forces in which the managers rightly recognize that fishermen and the wider public are not, in normal circumstances, ready to support the sort of measure which would achieve a genuine management improvement.

From a technocratic as opposed to democratic point of view, this is unfortunate, because it is only by decisive, radical, even revolutionary action that fisheries managers can hope to tackle the sort of problems which beset so many of the world's fisheries. Really effective curtailment of effort certainly requires radical action – and large public expenditure on buying out any surplus domestic effort. The introduction into this anarchic hunting industry of properly enforced and tradable property rights is pretty revolutionary – and can be made to work only if it is accompanied by substantial expenditure on an initial buy-out of any excessive domestic rights. These are the sort of measures needed – and they are just the sort of measures which, within an accountable political system, cannot be expected to be imposed on a domestic industry – and a domestic taxpayer – without a large measure of industry or public support.

I think is arguable that it is not even in the fishermen's interests, let alone those of the economy and the environment, that it is so difficult to get such measures introduced. The fishermen who obstruct the necessary improvements may, in the short term, be protecting something like the status quo but, in the longer run, nemesis does await. It might be better for the fishermen to engage in a constructive, realistic debate to devise a rational forward plan rather than impose on themselves as well as fisheries managers what can sometimes seem like a lifetime of crisis management.

THE EU DIMENSIONS

Let me now turn to the EU dimensions. The major factor here is really rather obvious. It is hard enough for one fisheries management authority to devise and implement a management regime. Where, however, the regime has to be negotiated between several different authorities, each with somewhat different interests and objectives, the difficulty of devising a satisfactory regime is hugely magnified. You are in just such a position in the EU, with several different member states and the Commission each wishing to have its particular views and interests accommodated.

Moreover, the problems are exacerbated by the fact that the regime finally agreed is not enforced consistently by a single authority but by each of the different member states subject to a not very effective invigilation by the Commission. All this is, I am afraid, a recipe for decisions which tend towards the lowest common denominator. Moreover, the institutions of the EU inevitably have the effect of reducing the responsibility and accountability of the individual participants. The Commission can always blame the Council, and any individual minister can always similarly blame his colleagues in the Council. As and when the North Sea cod collapses, I cannot see that any minister in any member state or the Fisheries Commissioner will have to resign. Lower levels of personal responsibility reduce the already low incentives for taking the necessary hard decisions.

Within the EU, the Commission is generally in a very powerful position to play member states off against each other and secure decisions which few, if any, member states really want. In the case of fisheries, the effectiveness of this manipulative power is weakened by the Commission's ultimate dependence on member states being willing and able to enforce the agreed rules. The Commission are also up against the fact that so far the management instruments which the Community have accepted for fisheries are simply not up to the job.

Hard decisions are, however, needed if the fisheries in member states' waters are to be properly conserved. There needs to be a real and substantial cutback in member states' fishing effort. The cutback has to be not just a paper cutback, but a cutback which is really enforced and is permanent. This may well require the introduction of new management instruments. This is all a very tall order. Bearing in mind all the agonies which went into the initial CFP settlement, and the political pain which would be involved in the necessary new measures, one cannot be optimistic about the prospects for such action.

THE UK's PROBLEMS

As far as our own national circumstances are concerned, the nature of the fisheries around our shores is such that, even if we were not in the EU, we would need to work with other countries to ensure the effective management and conservation of most of the stocks which interest us. Even when it comes to the management of our own fleet's fishing, we are somewhat constrained by our membership of the EU. The government did make an extremely brave and determined effort to get our own fleet into some sensible order through the policy of combined days-at-sea controls and decommissioning. However, that attempt was finally defeated by sheer politics, which is a pity because I think it could have led to a much less unstable and uncertain industry than we have now. However, the experience will, I think, make ministers very chary for the foreseeable future about adopting effective measures which apply only to the UK fleet.

Moreover, in pursuing any initiatives in the EU, ministers will always have to bear in mind that measures dependent on other member states' enforcement are likely to be unevenly applied, at least in some member states. We can of course, refine and tinker with our national quota management and inshore management regimes, but this does not seem to me to get anywhere near the heart of the problem.

There is, of course, the argument that the UK should withdraw from the CFP and rid ourselves of EU restraints. I am afraid I see this as a purely theoretical solution. As long as the EU exists and the UK is a member of it, I feel we will remain subject to *a* if not *the* CFP. The political and economic arguments against the UK trying to negotiate its way out of the CFP, while remaining part of the EU, seem to me just too big. This is not to say that the UK cannot and will not try to get the CFP improved.

However, I fear that the theoretical possibility of leaving the CFP actually somewhat harms the UK's efforts to improve the CFP in that it encourages the fishermen's and the public's all too human tendency to believe there is a painless solution to our over-fishing problems. If only, so the tale goes, we could eject quota hoppers from our fleet and all foreigners from our fishery limits, we could solve our fishing problem without doing anything unpleasant to our British fishermen. Even in theory, this would not be completely true: the impact of quota hoppers is far from uniform, and our indigenous fleet catches a large share of some hard-pressed stocks. However, the campaign for British withdrawal works against the adoption by the UK of a firm and effective strategy through the

CFP for tackling the serious over-capacity of our fleet, as w
that of other member states.

CONCLUSION

⌐I am afraid that, where the EU and UK fisheries are concerned, I
am not optimistic. I fear that the state of the stocks is likely to get
worse before the pressure for the necessary serious new measures
is strong enough to overcome all the political and other restraints.
We may even have to have a major stock collapse before there is
sufficient recognition by the public and the fishing industry of the
need for the industry itself to be cut back rigorously.

This is sad for all concerned. Crisis management is certainly no
fun for fisheries managers, nor I imagine for fishermen, particularly the more marginal operators. For fishermen, better management should mean that, while the size of the fleet and the number
of fishermen may be much reduced, the security and profitability
of those remaining would be much improved.

However, if we are going to achieve proper management of the
stocks around our shores, we must work with our European partners and we need a constructive input from the fishing industry.
Our own fishermen seem to be in danger of marginalizing themselves by simply being too negative and unrealistic. If fisheries
managers cannot find within the industry effective and responsible
interlocutors who are prepared to face up to the practical realities,
then there is a real danger that either the managers will give up,
making savings on the enormous current expenditure on fisheries
support, administration, enforcement and research or, especially if
the environmental interest in fisheries continues to increase, they
may concentrate on working with those interests who do recognize
the need for change so as to conserve the stocks.

However, in these remarks I do not want to get too deep into
the specifics of our particular fisheries problems. The general problem
I would like to draw out is that the lack of success of so many
fisheries managers seems to me to flow from the fact that economic, environmental and political arguments for effective measures
tend all too often to be counterbalanced by the economic and political arguments against. This is partly the result of the peculiar
economic and legal framework under which most sea fisheries

operate. A change in that framework, for example to instruments such as transferable effort quotas, could make a massive difference, but achieving such a radical change runs up against the fundamental political obstacle. Thus, although I am sure that, even here and within the EU, change will come, I fear that it is more likely to be driven by burgeoning or recurrent crisis than by more measured debate. To respond to the question posed in the title that I did not choose, that will *not* be just because fisheries managers are hopeless.

3 The Role of the European Parliament in the Development of the Common Fisheries Policy[1]

David Steel

INTRODUCTION

⌐The European Parliament is the only forum for the public discussion of fisheries legislation at the European level. At the national level there is generally no systematic scrutiny of European fisheries policy proposals; at the Council of Ministers level, discussion is conducted behind closed doors. The role of the Parliament is, therefore, essentially to ensure that legislation is properly discussed and to form a view on the basis of the arguments put forward. Its influence is determined by the way it plays this role.

While the Commission and Council may or may not take on board the opinion of the European Parliament – although the evidence suggests that the Commission does adopt a significant proportion of Parliamentary amendments – it is often said that the Commission is reluctant to take up a matter in the Council if it has not been aired in the European Parliament.

At the founding of the European Community, the European Parliament was a nominated assembly and fisheries was hardly a primary consideration of the framers of the Treaty of Rome. The Treaty of Rome in Article 38 mentions fish only in the context of its definition of agricultural products, namely 'products of the soil, of stock farming and of fisheries and products of first stage processing directly related to those products'. Article 43 of the Treaty stipulates that the Council is to legislate on the basis of proposals from the Commission and after consulting the European Parliament. As the European Parliament has developed and as the volume of fisheries policy has grown over the years so Parliament has sought, *inter alia*, to increase its democratic role with regard to the developing Common Fisheries Policy.

33

Furthermore, the Parliament can produce reports under its own initiative, some of which have contributed significantly to policy development. Members can also table both written and oral questions to the European Commission and the Council of Ministers. From time to time fisheries are also the subject of urgent resolutions and debates. Parliament's Committees hold hearings, they arrange for interest groups to present their case, and the MEPs can of course take up constituency or interest group points with relevant Commissioners and officials. By bringing a matter to the attention of Parliament a member can also 'europeanize' an issue and thus seek European support for a matter which may be peculiar to his or her own area.

In the UK recently, debate has focused largely on the subject of fish resources and who should manage them. Day-to-day fisheries policy of course covers a range of policy areas:

• conservation and management of the resources;
• market policy;
• monitoring and control policy;
• policy on fisheries agreements with third countries;
• structural policy;
• research policy.

This chapter will seek to focus on a number of issues within these policy areas to illustrate the role of the European Parliament.

THE FIRST LEGISLATURE

Over the decade from the enlargement of the Community to the settlement of January 1983 fisheries policy was a focus of community disharmony centred on the failure of the Council of Ministers to agree a conservation policy. Naturally, in this period fisheries industry representatives focused their attention and limited financial means on the Council of Ministers. The appointed European Parliament, which had given its opinion on the Commission proposals in 1977, was of little relevance.

In 1979, however, a new democratically elected Parliament was set up. In November, a fisheries working group was established with Robert Battersby, MEP for Humberside, elected chairman. Over the Parliament's first five years, the 22 members of the working group established its credibility as an interlocutor with the Commission with a series of reports on the development of a common policy, control and surveillance, the Mediterranean, social aspects of fisheries policy and the development of aquaculture. The Parliament was successful in achieving a separation between agriculture and fisheries in the Community budget, thus allowing a separately funded fisheries policy to develop. During this period, Parliament can take the credit for establishing a Community fisheries inspectorate – a major step forward in Community competence, even if the results to date have been limited, for ensuring a public debate and understanding of the questions related to the further enlargement of the Community, and for focusing attention on the plight of the Atlantic salmon – leading to the formation of the international North Atlantic Salmon Conservation Organization (NASCO). It is a mark of the speed of the development of the farmed salmon industry that the focus over the last ten years has shifted so dramatically from the problems of wild to farmed salmon. The Parliament also established itself as a voice for the peripheral areas dependent on fishing, again allowing for a solidarity and commonality of views between very disparate regions which nevertheless shared the common features of remoteness. Thus the particular case of the Shetlands has been both heard and accepted at the Community level.

This commonality of interests has been of significance to the UK more recently. Without such understanding and support, it is unlikely that the UK would have obtained special EU support (objective 1 status) for the Highlands and Islands of Scotland. And solidarity with Scotland and Ireland's remoter areas is well reflected in concern and support for the fish farming sector.

In particular, Parliament was able to stand back from the day-to-day management of the policy and point the way forward in several policy areas. Slowly, Parliament began to show itself as a bridge between the fishing industry at the local level and the Community institutions.

THE SECOND LEGISLATURE

During the second legislature from 1984 to 1989, fisheries became a full subcommittee of the European Parliament's Agriculture Committee.

The main reports of the Committee covered: the fisheries sector and enlargement, structural policy, research, fisheries agreements with third countries, the estimation and management of fish stocks, the common organization of the market and the fish processing industry. At the end of the period, the Committee produced a report in the name of Parliament's French Vice President, Nicole Pery, entitled 'Achievements and prospects of Europe's fisheries policy' (Pery, 1988).

The accession of Spain and Portugal in 1986, which doubled the number of Community fishermen and increased the tonnage of the fleet by some 50 per cent, also increased the political interest in fisheries in the European Parliament with MEPs from the new member states coming to play an active role in the Fisheries Committee. While the arrival of Spain and Portugal created new internal tensions, particularly with regard to access to North East Atlantic waters, it greatly increased interest in areas such as access to third-country waters, which is of particular importance to the Spanish distant-water fleet, structural policy and the common organization of the market. Spanish MEP Miguel Arias Cañete, who became Chairman of the Fisheries Committee in 1994, has also played an important role in pursuing fisheries questions through the budgetary procedure where Parliament and the Council are the two arms of the budget authority, and in establishing separate budgetary titles for fishing.

During the 1984 to 1989 period, the Parliament was probably most influential in the field of structural policy. One commentator noted the extent to which the new legislation agreed at the end of 1986 on structural policy took up proposals made by Parliament both in an own initiative report prior to the publication of the Commission proposal and in its report on the Commission's legislative proposals which were also largely accepted by the Commission in a revised text to its own proposal (de Feo, 1987). The 1986 legislation (Regulation 4028/86) provided the Community with substantial funding of 800 million ecu for the period 1987–91, not only for vessel modernization and new building but also significant sums for aquaculture, exploratory fishing and joint ventures, as well as limited aid for port facilities, market initiatives and other measures.

De Feo also reminds us of the Parliament's budgetary powers in this respect by inscribing a sum of 127 million ecu of expenditure in the 1987 budget, allowing the new measures to be financed.

THE THIRD LEGISLATURE

The 1989–94 Parliament focused, in particular, on important technical conservation measures, on the ongoing review of the Common Fisheries Policy and the impending integration of Spain and Portugal. There was also progress on the budget with the creation in 1990 of separate fisheries guarantee and guidance funds and the incorporation of fisheries in the structural funds agreed at the Edinburgh Summit in December 1992.

A significant feature of the period was the growing number of meetings between the fisheries subcommittee and fisheries representative organizations as well as the establishment of a series of seminars with Committee members and senior Commission officials, allowing an in-depth examination of important issues and developing contacts with the executive.

There was, too, a substantial increase in briefing material provided to MEPs on all the Commission proposals and EP reports. All these activities demonstrate not simply the growing role of the Parliament in policy formulation but of the movement away from heavy concentration on Council lobbying before final decisions towards focusing on the initial stages of legislation both at home and at the European level to ensure that relevant interests are taken care of *before* the publication of legislative proposals and throughout the legislative process. Let us look more closely at three examples.

CFP review

Following a lengthy period of consultation the Commission, in October 1992, presented a proposal (COM(92)387) to replace the basic regulation 170/83. The most significant aspect of the proposal was the continuation of the basic *acquis* of the CFP, namely:

- the principle of relative stability in the allocation of fishing opportunities;
- the system of reserved access in the 12-mile band;
- the arrangements for the Shetland Box.

The proposal also contained new measures for strengthening the CFP such as the introduction of a community licensing system, a proposal for multi-annual TACs, and the introduction of a new control system.

The European Parliament's involvement in the review of the CFP and consideration of future policy go back, as indicated, to 1988 with the presentation of Nicole Pery's far-reaching report on the achievements and prospects of Europe's fisheries policy. Significantly, this report stressed the importance of continuing the 12-mile limits and upholding the principle of relative stability. A report on the CFP prepared for Parliament by outside consultants stated: 'Although acting only in an advisory capacity Parliament has played a central role in every stage of the CFP review' (EP, 1994).

It is, therefore, interesting to note that, throughout the review process, Parliament maintained the principles set out in Pery's initial report. Thus, in its resolution of May 1992 on the Common Fisheries Policy and the adjustments to be made, Parliament stated:

> Notes that the basic principles of access to resources, ie the principles of relative stability and of protection of coastal waters within a twelve mile limit and 'boxes' notably the Shetland Box, will continue to be applied, as a majority of Parliament advocated, the sole aim of the planned adjustments being to increase their effectiveness.   (EP Minutes, 15/05/1992)

As far as the European Parliament is concerned, a striking difference between the basic regulation of 1983 (R170/83) and that of 1992 (R3760/92) is the fact that in the former Parliament was explicitly excluded from consultation, whereas in the latter regulation there are a significant number of references to consultation with the European Parliament. Indeed, with regard to fishing licences, where the Commission had proposed a management committee procedure, Council agreed to consultation with the Parliament pursuant to Article 43 of the Treaty.

Significantly, too, Parliament was subject to detailed briefing over this period from both the Scottish and English fishermen's federations. Not only did this information help to ensure that support for relative stability and access limitation was defended throughout the process but it gives a detailed insight into the thinking of those organizations at that time. One of the striking features is the emphasis on Community solutions, whether it be in the calls for the UK to participate in the decommissioning schemes or in the advocacy of Community-managed solutions. For example, not only did the NFFO envisage the possibility of greater control from the Commission over Multi-Annual Guidance Programmes (MAGPs) but a briefing paper of February 1992 stated with regard to enforcement: 'As such, the Federation takes the view that enforcement would be better co-ordinated and adopted by the EEC itself.' The industry also obviously felt that its lobbying had paid off to the extent that there was no threat to the six- and 12-mile limits. An NFFO note stated: 'The principle of 6- and 12-mile limits will be maintained in its existing format until 2002 and this policy has had wide support from NFFO and will not prove to be a controversial issue in the Fisheries Council.'

The Commission followed up the new basic regulation with a number of proposals as foreseen in the regulation. In the case of control measures the Commission had produced its original proposal in parallel with the basic regulation but proceeded to amend its proposal in the light of Parliament's report. In particular, the Commission agreed to the principle of comparable penalties for comparable offences in the member states – a subject which has exercised both the MEPs and the fisheries organizations. The Commission also agreed that an annual report should be drawn up on the controls implemented by the member states. The final control regulation adopted in October 1993 (R2847/93) made provision for the publication of an annual report on implementation of control measures, the first of which was published by the Commission in March 1996. While the regulation did not contain provisions for comparable penalties, as Parliament called for, a significant step forward, in response to Parliamentary appeals, was the strengthening of the Community inspectorate, allowing it to carry out checks without prior notice.

Technical conservation measures

In the case of so-called technical measures, the European Parliament would not normally be consulted on those measures required for the implementation of the Conservation Regulation 170/83. However, as Community fish stocks have declined, so 'technical measures' have assumed a greater political importance and we can see how Parliament has pressed successfully to have its say and how Parliament's opinion has helped to shape Community policy. In July 1990, for instance, the Commission published proposals to increase substantially the standard mesh size from 90 mm to 120 mm (COM(90)371). As the change was based on Regulation 170/83 the European Parliament would not be consulted.

However, at the first meeting of the fisheries subcommittee after the submission of the proposal British MEP Paul Howell drew Parliament's attention to the far-reaching effects of the Commission proposal and called for Parliament to be consulted. This was followed by a letter from British MEPs of all parties to President Delors and by a debate in the European Parliament calling again for consultation. Significantly at this debate, Commissioner Marin, the commissioner responsible for fisheries, backed Parliament's call for consultation (EP Debates, 12/12/90). The Council of Ministers, having failed to reach agreement on the proposals, belatedly agreed to consult the Parliament. In October 1991 Parliament adopted its opinion calling, *inter alia*, for an increase in mesh size to 100 mm with further increases on a step-by-step basis (EP Debates, 10/10/ 91). When the Council of Ministers met at the end of the month, agreement was reached very much along the lines of the solution proposed by the European Parliament. We shall examine further the role played by the Parliament in connection with large-scale drift netting for tuna and the renewed interest in technical measures reflected in the Commission's recent report (COM(95)669).

Integration of Spain and Portugal

December 1992 saw the presentation by the European Commission of its report on the application of the Act of Accession of Spain and Portugal in the fisheries sector (SEC(92)2340). While the Commission's report, required under the Act of Accession, recognized that the arrangements, subject to any amendments, were due to continue till 2002, it was sympathetic to Spanish concerns. It described the transitional period as being particularly long in comparison with the average duration of transitional arrangements. It was the Commission's view that fishing by Spanish and Portuguese vessels in the waters of the ten and vice versa must be examined not in the climate of 1983, when the basic conservation policy was in its infancy, but in the climate of 1992.

Under the accession arrangements there was a limit to the number of vessels which could fish in the waters of the ten and a reporting-in system with notification of catches on board. While the Commission noted that 'vessels do not always comply with these obligations' and expressed concern at the failure of the Spanish authorities to remove vessels involved in serious infringements from the periodic lists, it nevertheless concluded that the input management system, as it applied only to the fishermen of Spain and Portugal, led to a feeling of discrimination. Looking ahead, the Commission believed that there should be a general licensing system applicable to all Community fleets and the periodic lists should be converted into a general system of effort control.

Within the European Parliament, a report was prepared by a member from the Azores, Vasco Garcia (Garcia, 1993). While Garcia endorsed the need for change while other provisions would remain till 2002, his remarks concerning the Irish Box are of particular interest. The Commission was of the opinion that the Irish Box restrictions mentioned in Article 158 of the Accession Treaty and tuna restrictions mentioned in Article 351 'will be without effect as from 1 January 1996'. Although the Garcia report stated that the total exclusion would end on 1 January 1996, it called for the Commission to consider that the area be given the same 'sensitive area' status as the Shetland Box, the Azores, Madeira and the Canary Islands. In Parliament's debate Irish members in particular focused on Garcia's references to the Irish Box and called on the Commission to take the matter further. It was not, however, till October 1993 that the Commission produced a proposal, to allow the integration of the accession arrangements into the common policy with effect from January 1996, leaving little time for Parliament to give its formal opinion and Council to reach agreement by the agreed January 1994 deadline (COM(93)493). Parliament shared the British government and industry view that there should be no blank cheque for Spanish and Portuguese access. The text brought to the Parliament in December not only sought to simplify the text presented by the Commission but came out in favour of a conservation zone for the Irish Box (EP Minutes, 17/12/93).

The agreement eventually reached by the Council in May 1994 (R1275/94) provided for the Commission to present proposals by June 1994 to contain and control fishing effort in Western waters, ensuring that the new European Parliament also elected in June 1994 would be faced again with this question.

THE FOURTH LEGISLATURE

The fourth legislature, elected in June 1994, saw a further enhancing of the role of fisheries in the European Parliament with fisheries elevated to full committee status. Under the leadership of Miguel Arias Cañete, the Committee has sought to ensure that fisheries subjects are brought to the public's attention through debate both in Committee and in the Parliament's plenary sessions. The oral question with debate procedure has been used to tackle the Commission and Council of Ministers on subjects as varied as the fisheries agreement with Morocco, fishing in the North Atlantic Fisheries Organization (NAFO) area, the recurring crisis in the salmon sector and the Commission's attempts to push through veterinary inspection charges without Parliamentary consultation. Let us look closely at three issues: integration, fish farming and tuna drift netting.

Parliament returns to the integration question

In July 1994 the Commission produced two proposals, one establishing rules for access to certain Community fishing areas and resources, the other on control arrangements (COM(94)308 and COM(94)309). The former proposed a system of standard vessel days, the latter a system of monitoring vessel movements through a reporting-in and catch-reporting system. In the European Parliament, Spanish Socialist member Maria Izquierdo Rojo was appointed to present Parliament's opinion on the access arrangements and British Conservative member James Provan was appointed rapporteur on the control arrangements.

A majority of Council members rejected the Commission proposals at their meeting of 28 September 1995 and began to work on building a Presidency Compromise on the access arrangements, leaving the control arrangements on one side. The Council, under pressure to block enlargement if agreement were not reached, was also dependent on receiving Parliament's opinion.

While the Izquierdo draft report rejected the Commission's more bureaucratic proposals, it made no mention of the Irish Box and said nothing about relative stability and preventing increases in fishing effort. The Izquierdo report and the Commission proposal were rejected in the Fisheries Committee. In an effort to ensure that Parliament would give its opinion on time, a compromise text was agreed under Miguel Arias Cañete's chairmanship which, while it endorsed the declaration of the General Affairs Council that 'the fishing fleet of Spain and Portugal should be fully integrated into the CFP from 1 January 1996', underlined the need for respect of the *acquis communautaire*, especially the principles of relative stability and no increase in fishing effort, endorsement of a vessel list system in place of the complicated vessel days proposal and support for the Irish Box (EP Minutes, 19/12/1994). On receipt of Parliament's opinion, the Council was able to reach agreement on 22 December – just within the timetable it had set for itself. The ensuing regulation (R685/95) would hardly have contained the continuing limitation of access to the Irish Box if it had not been for the insistence of the European Parliament.

With regard to control measures, the European Parliament has insisted on many occasions that such measures have to be fair and have to be enforced equally throughout the Community. Control measures have to be credible, understandable and cost effective. It was James Provan's view that the Commission's control proposal answered none of these criteria.

Although the Council of Ministers sought an early opinion from Parliament on the proposal, Provan persuaded the Fisheries Committee that Parliament should not give an opinion on a measure which would never be accepted by the Council of Ministers. Parliament was not prepared to rush out an opinion which could simply be ignored.

In a working document Provan set out his concern that the proposal would add greatly to the administrative burden on fishermen without improving control. He wanted to see more effective controls with less burden on fishermen. Following the Council agreement of December 1994 the Commission eventually came forward with a modified proposal (COM(95)256). While many fishermen opposed both any increase in reporting-in requirements and the further use of satellite technology, Parliament supported Provan's view that satellites and black boxes were the only way of strengthening control, while at the same time lessening bureaucratic burdens, particularly on the largest 4 per cent of vessels which are responsible for 35 per cent of the Community catch (EP Debates, 21/09/1995 and Provan, 1995). Thus, while the European Parliament in this instance contributed to the rejection of over-bureaucratic proposals it helped to keep the issue of high technology control measures on the agenda.

Scottish and Irish salmon farming

As early as 1989 the Scottish and Irish salmon farming industries had asked the Commission to investigate allegations of dumping by Norwegian salmon exporters. When, in 1991, the Commission recommended the imposition of an 11 per cent tariff against Norway it received no support from member states other than the UK and Ireland. At a seminar with Commission officials, MEPs were told that not only did other member states not support action against Norway but even the UK and Ireland did not give their full support. This stimulated action by MEPs who arranged for the salmon growers to meet with officials of the private office (cabinet) of the then Trade Commissioner Frans Andriessen, with fisheries sector officials and with the Parliament's Fisheries Committee.

The growers expressed their concern to the Fisheries Committee that 'political considerations in the EC/EFTA discussions are working against salmon farmers. . . . The impression we have had is that these talks are seen as more important than our problems. . . .' Members took up their concerns and were able to help organize meetings with the relevant Commissioners. Arrangements were also made for the Secretary of State for Scotland, on a visit to Strasbourg, to meet Commissioners and press the case for the Scottish industry. Parliament maintained its pressure on the Commission, and the Fisheries Committee, in particular, continued to focus on the market situation. Thus, when the subject was debated in the European Parliament in November 1991 (EP Debates, 22/11/1991), MEPs were able to welcome the Commission's decision to introduce minimum import prices.

The contacts built up in 1991 demonstrated the role which the Parliament could play in bringing the problems of a particular industry to the Commission's attention, especially when wider concerns may have militated against such an industry, which was only important to the peripheral parts of two member states. The exercise also illustrated the effects of a united lobbying effort by the industry.

Over the years, MEPs have maintained their concern for the salmon farming industry with Parliamentary Questions and a resolution in November 1993 as the market price fell again. In January 1994 the European Commission intervened again and in February 1994 the salmon growers lodged a complaint against Norwegian state aids. MEPs also took the matter up directly with the Norwegian government in the course of Norway's negotiations to join the Community.

At the Fisheries Committee meeting in November 1995 members pressed the Commission representative on the sudden and severe decline in the price of salmon on the Community market. The Committee decided to table an oral question for debate and, because of the enhanced status of the Fisheries Committee and the widespread concern, the item was put on Parliament's agenda at the next plenary session in December when a resolution was adopted calling on the Commission to take immediate action to reimpose a minimum import price on salmon. It is no coincidence that the Commission decided the same day to adopt a regulation making the release for free circulation of salmon of European Economic Area (EEA) origin conditional upon observance of a floor price (Commission Regulation, 2907/95).

This time the minimum import price did not have an immediate effect and the industry came back to the Parliament. It became clear from discussions with the new Fisheries Commissioner, Emma Bonino, that the Commission was reluctant to take measures to impose quantitative limitations on Norwegian salmon. However, the Commission would assist with an anti-dumping, anti-subsidy complaint. Following a further Parliamentary resolution the Fisheries Committee organized a hearing on the subject which was attended by the Scottish and Irish industries and a Norwegian delegation headed by their Ambassador to the European Community. The upshot of the meeting was that the industry decided to pursue a further complaint against Norway.

Large-scale drift netting for tuna

The Commission proposal of April 1994 (COM(94)0131) for the phasing out of large-scale drift netting led to some of the fiercest arguments over fisheries policy in the Parliament. In 1989 a resolution was tabled calling for Community legislation to ban high seas drift netting in its waters and to support international efforts to regulate purse seine and drift net fishing for tuna (EP Resolution, B3-0012/89). This resolution was followed up with reports on both drift netting and purse seining.

When Parliament considered the Commission proposal to limit drift netting for tuna in Community waters to less than 2.5 km it was noticeable that while UK members strongly opposed the fishery and were lobbied against drift netting by environmental and animal welfare organizations, the French authorities drew on local 'green' organizations such as 'Robin des Bois' to support French drift net fishing. At the end of the day Parliament's endorsement of the proposal helped to smooth the way for agreement in Council (EP Debates, 10/10/1991).

However, the campaign against drift netting grew, and in

(December 1993 Parliament adopted a resolution calling for a total ban on the use of drift nets (EP Minutes, 17/12/1993). In April 1994 the Commission came forward with a proposal to phase out drift nets altogether (COM(94)0131). The European Parliament decided to await the elections due in June before preparing its report – rejecting a Council request for an urgent opinion at its May plenary session. At the first meeting of the Fisheries Committee the new Spanish member Carmen Fraga was appointed to prepare Parliament's opinion. In September the Committee voted by 10 votes to 8 to back the rapporteur's call for an immediate ban on drift net fishing. The Fraga report was subsequently adopted by Parliament as a whole (EP Minutes, 29/09/94 and Fraga, 1994).

To date, the Council of Ministers has failed to reach agreement on the Commission proposal. The member states and the Commission have been involved in expensive monitoring exercises and the number of vessels involved has declined – confirming the view of fishermen that they could not fish viably with nets of 2.5 km. While Parliament had pressed successfully for the Commission to introduce legislation – siding in this case with the majority of environmentalists and the traditional pole and line fishermen – there could be no agreement in the Council of Ministers without the required majority.

CONCLUSION

The examples given above show how the European Parliament has played its role in ensuring that the Commission's reports and proposals are properly debated. Parliament's opinion may not necessarily have been adopted by the Council of Ministers and the European Commission but, over the years, increasing attention has been paid to the view expressed by Parliament. The evidence suggests, too, that where the British industry, MEPs and government have worked along the same lines, results have been achieved which could not necessarily have been anticipated at the outset. Equally, the current divisions in the British industry may have made it more difficult for industry representatives to focus on European Parliamentary initiatives.

At the individual level, MEPs, through questioning the Commission and Council, through research, through their contacts with the industry and through their responsibility for Parliamentary reports, have been able to build up a respect for their understanding of the fishing industry and have thus been able to bring their influence to bear on Commission, on government and on the fishing industry.

Over the years the Fisheries Committee of the European Parliament has focused on the Commission's reports and proposals. Sometimes, as in the review of the CFP, it has focused on fundamental aspects of the policy, in other cases it has considered the many more technical proposals. The Parliament, too, under its own initiative has, in certain cases, anticipated developments and sought to ensure public debate ahead of important proposals. It has also voiced its concern over matters where it has not been formally consulted.

At the institutional level, the Fisheries Committee has sought to strengthen Parliament's formal role in fisheries policy-making. In particular, the Committee, taking the lead from the opinions prepared by Brigitte Langenhagen, MEP, has wanted to see a separate title for fisheries in the Treaty to further distinguish it from agriculture; secondly, the Committee has wanted to strengthen Parliament's role in decision-making by applying the co-decision procedure to the basic regulations of the Common Fisheries Policy – i.e. to give Parliament greater authority *vis-à-vis* the Commission and Council of Ministers. Finally the Committee has called for the assent procedure to be applied to all fisheries agreements, thus making the go-ahead for the implementation of agreements dependent on a vote in Parliament.

In its report adopted in May 1995 Parliament called for the general use of the co-decision procedure and the use of the assent procedure for all international agreements. It also stated:

> Powers in the field of fisheries need to be dealt with independently of those in the field of agriculture. The Common Fisheries Policy should be re-examined in accordance with the founding principles underlying the institution of the common policy, i.e. conservation and relative stability.   (EP Minutes, 17/05/1995)

Before the opening of the IGC process at Turin on 29 March 1996, Parliament provided a further opinion where, as well as stressing the need for a simplification and strengthening of the legislative procedure, it proposed a specific fisheries heading. Here it was stated clearly that a separate fisheries title should be included in the Treaty and that the assent procedure should be applied to all international fisheries agreements (EP Minutes, 13/03/1996).

At the fisheries policy level, Parliament is currently considering the future multi-annual guidance programmes. It is working on the

Commission's report on the implementation of technical measures in the Common Fisheries Policy and will report on the follow-up proposals. It will also build on the recent Commission report on monitoring the Common Fisheries Policy (COM(96)100). In both these areas Parliament's rapporteurs have already focused on the lack of confidence in the member states' ability to apply the regulations equitably throughout the European Community.

Industry's concerns have been voiced more generally in the report by Nicole Pery, MEP, on the crisis in the fishing industry (Pery, 1996). Here again, Parliament has stressed the need to act on the concerns of those connected with the industry by involving them more closely in policy-making to restore their confidence in the European Union's management of the CFP. While the report ranges widely over the Common Fisheries Policy from problems in the common organization of the market to calls on the Commission to strengthen structural measures, it is interesting to note that it takes up a number of the points which are currently exercising the industry in the UK but also are of concern elsewhere in the Community:

- the need to improve resource management with particular regard to technical measures;
- the need for a greater involvement of those fishing in individual sea areas in resource management;
- the need to tackle further enlargement so that the adjustments of the CFP to resolve the fisheries crisis do not conflict with enlargement;
- the need for the Commission to draw up conditions which can be applied by member states to limit quota hopping.

Significantly, however, while making many proposals for strengthening the CFP, there is no questioning of the need for a common policy (EP Minutes, 20/06/1996).

Parliament is also looking more closely at other fundamental aspects of the CFP. Against the background of increasing concern about fisheries agreements, particularly with third world countries, a report will be prepared on the future of fisheries agreements. A meeting has already been held with the industry and Commission officials to focus on fisheries agreements. Work has also begun on looking ahead to the year 2002 and beyond. Hearings have been planned for autumn 1996 to begin to look at comparative schemes for managing fish stocks and to ask whether scientific research is meeting the needs of the industry. As with fisheries agreements a rapporteur has been appointed to consider the CFP and to look ahead to the 2002 review and beyond.

The Fisheries Committee of the European Parliament has set in place the framework for the debate at the European level on the fisheries policy requirements of the new century. Past experience suggests that the European Parliament can make a useful contribution to the debate and will be increasingly well placed to scrutinize Commission proposals in the months and years ahead.

NOTE

1. The views expressed in this paper are those of the author alone.

BIBLIOGRAPHY

Commission of the European Communities

1990 *Proposal for a Council Regulation (EEC) Amending for the Tenth Time Regulation (EEC) No. 3094/86 Laying Down Certain Technical Measures for the Conservation of Fishery Resources.* COM(90)371, 31 July 1990.

1990 *Proposal for a Council Regulation (EEC) Amending for the Eleventh Time Regulation (EEC) No. 3094/86 Laying Down Certain Technical Measures for the Conservation of Fishery Resources.* COM(90)610, 7 December 1990.

1992 *Proposal for a Council Regulation (EEC) Establishing a Community System for Fisheries and Aquaculture.* COM(92)387, 6 October 1992.

1992 *Report by the Commission to the Council and Parliament on the Application of the Act of Accession of Spain and Portugal in the Fisheries Sector.* SEC (92)2340, 23 December 1992.

1993 *Proposal for a Council Regulation (EEC) on Adjustments to the Fisheries Arrangements Provided for in the Act of Accession of Spain and Portugal.* COM(93)493, 13 October 1993.

1994 *Proposal for a Council Regulation (EC) Amending for the Sixteenth Time Regulation EEC No. 3094/86 Laying Down Certain Technical Measures for the Conservation of Fishery Resources.* COM(94)131, 8 April 1994.

1994 *Proposal for a Council Regulation (EC) Establishing Rules for Access to Certain Community Fishing Areas and Resources.* COM(94)308, 13 July 1994.

1994 *Proposal for a Council Regulation (EC) Amending Regulation (EEC)*

No. 2847/93 of 12 October 1993 Establishing a Control System Applicable to the Common Fisheries Policy. COM(94)309, 13 July 1994.

1995 *Proposal for a Council Regulation (EC) Amending Regulation No. 2847/93 Establishing a Control System Applicable to the Common Fisheries Policy.* COM(95)0256, 12 June 1995.

1995 *Implementation of Technical Measures in the Common Fisheries Policy – Communication from the Commission.* COM(95)669, 15 December 1995.

1995 *Commission Regulation (EC) No. 2907/95 of 15 December 1995 Making the Release for Free Circulation of Salmon of EEA Origin Conditional upon Observance of a Floor Price.* OJ No. L304, 16 December 1995.

1996 *Monitoring the Common Fisheries Policy – Commission Report.* COM(96)100, 18 March 1996.

Council of the European Communities

1983 *Council Regulation (EEC) No. 170/83 of 25 January 1983 Establishing a Community System for the Conservation and Management of Fishery Resources.* OJ No. L24, 27 January 1983.

1986 *Council Regulation (EEC) 4028/86 of 18 December 1986 on Community Measures to Improve and Adapt Structures in the Fisheries and Aquaculture Sector.* OJ No. L376, 31 December 1986.

1992 *Council Regulation (EEC) No. 345/92 of 27 January 1992 Amending for the Eleventh Time Regulation (EEC) No. 3094/86 Laying Down Certain Technical Measures for the Conservation of Fishery Resources.* OJ No. L42, 18 February 1992.

1992 *Council Regulation (EEC) No. 3760/92 of 20 December 1992 Establishing a Community System for Fisheries and Aquaculture.* OJ No. L389, 31 December 1992.

1993 *Council Regulation (EEC) No. 2847/93 of 12 October 1993 Establishing a Control System Applicable to the Common Fisheries Policy.* OJ No. L261, 20 October 1993.

1994 *Council Regulation (EC) No. 1275/94 of 30 May 1994 on Adjustments to the Arrangements in the Fisheries Chapters of the Act of Accession of Spain and Portugal.* OJ No. L140, 3 June 1994.

1995 *Council Regulation (EC) No. 685/95 on the Management of the Fishing Effort Relating to Certain Community Fishing Areas and Resources.* OJ No. L71, 31 March 1995.

Miscellaneous

de Feo, A. (1987) 'Les progrès de l'Europe bleue et la participation du Parlement Européen', *Revue du Marché Commun*, No. 309.

European Parliament (1994) *Manual of the Common Fisheries Policy.*

Fraga, C. (1994) *Report on the Proposal to Amend Regulation 3094/86 Laying Down Certain Technical Measures for the Conservation of Fisheries Resources.* DOC A4-0009/94, European Parliament.

Garcia, V. (1993) *Report on the Commission Proposal for a Council Regu-*

lation (EEC) on Adjustments to the Fisheries Agreements Provided for in the Act of Accession of Spain and Portugal.* DOC A3-0433/93, European Parliament.

Pery, N. (1988) *Report on the Achievements and Prospects of Europe's Fisheries Policy (Blue Europe).* DOC A2-0319/88, European Parliament.

Pery, N. (1996) *Report on the Report from the Commission to the Council and the European Parliament on the Crisis in the Community's Fishing Industry.* DOC A4-0189/96, European Parliament.

Provan, J. (1995) *Report on the Proposal for a Council Regulation No. 2847/93 Establishing a Control System Applicable to the Common Fisheries Policy.* DOC A4-0210/95, European Parliament.

4 The Determinants of Fishing Policy: a Comparison of British and French Policies
Mireille Thom

INTRODUCTION

The crisis in fisheries management has generated interest in the structural arrangements within which the fishing policy process evolves. This interest generally stems from a desire to improve the environment of decision-making in order to enhance legitimacy through greater user participation in policy formulation and implementation. Comparing fishing policies in the United Kingdom (UK) and France over the last 13 years, it is argued in this chapter that, while structures are a crucial part of fisheries management, inherent sectoral constraints are the main determinants of fishing policy outcomes. In this context, any approach other than global will fail to cure the ills associated with the exploitation of fish resources.

THE DETERMINANTS OF FISHING POLICIES

Studies show that in fishing, as in other industrial sectors, policies emerge from patterns of interaction between government and industry representatives. For reasons of democracy, legitimacy and a need for expert contribution, various groups have become involved in decision-making. As a result the policy process has become segmented and sectorized (Jobert and Muller, 1987). The linkages and interactions between the various players, which may include interests beyond those of the industry concerned, are generally conceptualized as a policy network (Jordan, 1990). Within this network there is usually a core, or policy community, with a restricted number of participants, able to exercise control over its access (Marsh and Rhodes,

52

1992: 251). Depending on the involvement of the industry in the process, policy networks and communities have been linked to pluralist, corporatist and elitist concepts of policy-making (Marsh, 1995). Nonetheless, there is a consensus that these structures and the resulting patterns of interaction influence policy (Marsh and Rhodes, 1992; Marsh, 1995). In this context, fishing policy differences should emerge between the UK and France where traditional state/society relationships differ (Shackleton, 1986), policy styles show variations (Jordan and Richardson, 1982; Hayward, 1982) and government action in the economic and industrial sectors ranges from interventionism in France to laissez-faire in the UK (Atkinson and Coleman, 1989).

In contrast to this concept, a sectoral policy approach would predict similar policy outcomes in the two countries. According to this concept, policies and politics show '*differentiation* within individual countries across sectors and *convergence* across nations within sectors' (Freeman, 1985: 486; emphases in the original). In fisheries management policies would show similarities due to the intrinsic constraints of the sector. These include the characteristics of the resource, its ownership pattern, the high level of externalities on states, the multi-levelled types of pressure, trade agreements, the fragmentation of the industry and its lack of economic and political weight. Thus, fishing policies would be determined by these sectoral imperatives regardless of the national environment which produced them.

The UK and France offer a good comparative case study to test these two conceptual approaches. Both countries share the same sectoral and institutional constraints as members of the European Union (EU). They also enjoy a similar degree of autonomy in the implementation of the Common Fisheries Policy (CFP) regulations. After a brief examination of their respective policy environments, UK and French policies on structures, quota management and control of fishing activities will be examined since the inception of the twenty-year CFP in 1983.

THE POLICY ENVIRONMENT IN THE UK AND FRANCE

The administration

In both countries, fisheries is the responsibility of a central government ministry. Reflecting the territorial administration of the UK, the ministry in London (MAFF) has three junior partners in Scotland, Northern Ireland and Wales. While all have a policy input, MAFF ultimately makes the decisions. France's traditional paternalistic attitude to the fishing industry is still reflected in the sector's separate welfare system which is managed, *inter alia*, by a specific administration, the *Affaires Maritimes*, also responsible for enforcement of regulations, overseeing the allocation of public funds and recording and transmitting the industry's grievances. There is no such administration in the UK; instead, 18 local inspectorates control landings and monitor quota uptake. The French dirigiste approach is illustrated in the high degree of intervention and planning in the shape and nature of the fishing fleet (Meuriot, 1986) and in the various marketing and management structures and commissions set up over the years by the authorities. In Britain, on the other hand, the attitude has always been one of laissez-faire and *ad hoc* support. Aid has traditionally been targeted at specific fleets but with the simple aim of seeing the recipients through a bad patch (Thom, 1993).

Fishermen's organizations

British representation of fishing interests mirrors both the territorial arrangements and the pluralistic policy style generally associated with the UK (Jordan and Richardson, 1982). Many local voluntary associations have been created to give a voice to varied and, often, competing interests. The majority of associations are affiliated to the National Federation of Fishermen's Organizations (NFFO) in England and Wales and the Scottish Fishermen's Federation (SFF) in Scotland, while Northern Ireland retains two FOs. The federations and officials from the various departments form policy communities which have, hitherto, managed to remain closed to outsiders such as environmental groups. Consultation papers are regularly issued and officials and representatives meet formally and informally and exchange views by letter and over the telephone. Undoubtedly, representation along territorial lines weakens the industry position as there are some fundamental disagreements between the two federations. These divisions are exploited by civil servants who can adopt a UK-wide approach. Industry involvement in implementation is limited to the management of quotas. This task is undertaken by commercial structures, the producer organizations (POs), set up with EC backing to promote the interests of producers. In addition, 12 Sea Fisheries Committees (SFCs) manage the six-mile inshore waters in England and Wales, but interests beyond those of the industry are represented on these SFCs.

In contrast, French representation is organized through a national, corporatist and hierarchical structure, set up by the French Parliament to represent equally all component branches of the fishing sector from catchers to processors and merchants. Membership of the semi-public organization, the *Comité National des Pêches Maritimes et des Cultures Marines* (CCPM until 1993), is compulsory. This structure is strengthened by ten regional and 39 local committees (CLPMs). The CNPMCM's role is wider than representation, as it has the power to regulate coastal fisheries as well as manage specific stocks. The various committees must work closely with the field administrative structure, the *Affaires Maritimes* which is also represented at regional and local levels. In effect, consultation and policy-making have been institutionalized (Hennequin, 1989), but the prime role of the structure, which remains the 'creature of the authorities' (Jagot, 1996), is as a buffer between government and industry and industry and government (Rabot, 1993). Overall, French PO and committee involvement in policy implementation has been quite limited and subordinated to the supervision of the administration. Strong differences divide the French industry, the main cleavage operating between the industrial and artisanal sectors, with both having their own POs and associated organizations.[1]

To what extent do these differences in tradition, policy approaches and structures lead to different policy outcomes in the two countries? The next sections examine and compare the implementation of the CFP's three policy strands from 1983 onwards.

THE STRUCTURAL POLICY

In both countries after the war, public funds were made available to investors. In the UK, there was no concerted attempt to use structural aid to shape the fleet. It was felt that investors were best placed to make the right decisions regarding fleet development (Cmnd, 1961: 682). In France, multi-annual plans were devised to target specific fisheries and encourage various management structures and techniques borrowed from the industrial sector (Meuriot, 1986). The EC adopted a similar interventionist line, using the CFP's structural policy initially to develop and modernize the various national fleets before turning it into an instrument of fleet and fishing effort reduction. Three EC Multi-Annual Guidance Programmes (MAGPs) were created from January 1983 to December 1996. The Community along with national authorities co-financed selected projects both for the decommissioning of older units and the renewal of the fleets. Targets were set and agreed by governments and the Commission.

Implementing the 1983–87 MAGP

Both the UK and France had to ensure that fleet development was balanced out by equivalent capacity withdrawal. Both needed to scrap their large vessels and build medium and smaller units to fish in EC waters. Following consultation, the UK decided on a scrap and build policy by providing funds for a decommissioning scheme and grants for fleet renewal. In keeping with tradition and the market philosophy of the government, decisions regarding fleet development were left to investors. Industry calls for the setting up of a comprehensive licensing scheme and the creation of a national licensing agency were ignored (SFF, 1983). Instead, a partial licence system was established to throw a ring around the *number* of vessels prosecuting stocks under pressure at their 1984 level. The 1967 Sea Fish (Conservation) Act had enabled the government to take measures to implement the UK's first MAGP and to meet its targets. The UK was well placed to enter its second.

France, in contrast, decided on policy inertia. Industrial vessels were being laid off for lack of fishing opportunities, thus freeing tonnage and capacity units. EC grant aid for fleet renewal was matched but no attempts were made to licence vessels or fund decommissioning. In any event, France could not have acted quickly as the existing legislative framework had not been adapted to the new European environment. The necessary enabling law was only passed in May 1985 and the related secondary legislation had to wait until the early 1990s. The focus was exclusively on developing all branches of the sector. Nonetheless, the EC Commission judged that, though some vigilance was required, France, like the UK, had met her MAGP targets (EC Commission, 1986b: 28).

The technical and political effects of structural measures take years to become visible. While the overall fleet capacity had remained stable, fishing effort was relentlessly increasing due to the much improved performance of new technology. The UK government was severely censured by the Public Accounts Committee for its decommissioning scheme, described as 'grossly expensive for what it (had) achieved' (PAC, 1988: v). In addition, the licensing of vessels over 10 metres had led to the proliferation of 'rule beaters' – high-powered units marginally under 10 metres (Rodgers, 1989: 2). Finally, licences, allocated free by the government, had now acquired a value estimated by the Scottish Office at between £5000 and £10 000 (NAO, 1987: 10).

Second MAGP (1987–91)

Instead of the modest cuts requested in the second generation of MAGPs, capacity and tonnage continued to increase with dramatic effects on resources. By the late 1980s, the Commission, angry with several member states, ignoring their MAGP objectives, suspended grant aid for new vessels. Responding to industry calls, the UK sought cuts by attempting to expel foreign owned vessels from its register. However, the offending provision in the Merchant Shipping Act had to be suspended by Parliament when it was found to contravene European law (HC, 1989: 993–1015). This signalled a deterioration in the relationships between the UK government and Brussels and between the government and the fishing industry. The authorities were to remain deaf to calls, in response to successive consultation papers, for decommissioning measures, a tightening of the licensing rules and a greater use of technical measures. The policy communities were active and the government made some adjustments but, overall, the industry failed to influence the government's thinking on structures. The Scottish Office Fisheries Department was also engaged in a struggle with MAFF in favour of decommissioning but had to wait until the eve of the 1992 general election to win the argument. By that time, a few months away from the end of MAGP II, the UK fleet was some 19 per cent over capacity.

Again France's response contrasted with the UK's. Despite a change of government in 1986, nothing happened for another two years. The authorities simply sought to reduce the requested cuts and continued to encourage investment. Not surprisingly, instead of a projected 2.7 per cent cut in 1987, fleet capacity leaped by 7 per cent. The freezing of aid programmes alarmed the industry who knew that inertia was no longer an option. Aware of the impending presidential election in May 1988, the government adopted

a wait-and-see attitude and failed to make any proposals. The new
minister convened a meeting involving the Fisheries Secretary, representatives from the legal services, the minister's cabinet, the CCPM,
the UAPF,[2] the Cooperation[3] and the *Affaires Maritimes*. The minister explained that action was inevitable to ensure the return of
aid schemes. Despite France's deep-seated hostility to licences, the
authorities made it clear that they had to make the delivery of a
permit (PME)[4] compulsory before introduction of new capacity.
CCPM minutes of the meeting show that the industry was resigned
and, while concern was expressed over the *principle* of the scheme,
participants concentrated on its modalities. New capacity could not
be introduced without a withdrawal greater than any proposed
additions. Unlike the UK licences, the French permit cannot be
transferred. Since the legal instruments requiring such a permit
were not available, the CCPM's regulatory power had to be used.
Yet, while the CCPM processed applications, effectively the administration, which had drafted the scheme, ran it. This measure
was not sufficient to stem the increase in capacity and a decommissioning scheme was consequently set up in March 1991. This
plan, which contained social measures to help redundant crews,
successfully removed enough capacity to ensure that France had
met her MAGP II objectives.

Though different, the UK and French licensing and permit schemes
led to speculation over licences and kilowatts of engine power. Thus,
the French kilowatt was worth some £270 in 1990 (Weber and Antona,
1990: 7). In both countries, decommissioning financed the withdrawal of mainly old vessels which would have left the industry
anyway and whose impact on the resources was negligible (Durand
et al., 1992; Thom, 1993).

Third MAGP (1992–96)

Having used the structural policy to legitimize the CFP regime
through grant aid, the Commission now saw it as a tool to reduce
fishing effort. The Commission proposed such drastic cuts for the
third generation of MAGPs that member states refused to endorse
them and implementation had to be delayed for a year. A uniform
2 per cent cut was requested on the 1991 objectives for 1992. MAGP
III spelt out the level of reduction required in the newly set fleet
categories. Effort reductions could now form part of MAGP III.
Finally, the three policy strands – conservation, structures and control

– were brought together in the 1992 CFP basic regulation and a comprehensive control regime was set up in 1993.

Spurred on by political considerations – the forthcoming presidency of the EC, the CFP review and a looming general election – the UK government relented and proposed a £25 million decommissioning scheme effective over three years. It was, however, made conditional on the setting up of an effort control scheme. Decommissioning aid was not a right for departing skippers; instead candidates had to submit tenders to the authorities who selected the bids. POs were encouraged to buy licences from those leaving, resulting in the industry financing decommissioning. The industry was furious. It wanted more decommissioning funding and fought to change the conditions but, as the Fisheries Minister told the House of Commons, 'the link between decommissioning, effort control and all of the other measures (was) not negotiable' (HC, 1992: 57). Legislation regarding activity control was rushed through Parliament in summer 1992, but was abandoned in 1995 due to administrative difficulties, industry hostility and legal proceedings by the NFFO. A further £28 million was subsequently added to the initial £25 million budget in January 1995 to avoid defeat in a House of Commons vote. As MAGP III drew to a close, decommissioning had only removed 8.7 per cent of UK fleet capacity (Fisheries Department, 20 August 1996). Other measures such as penalties on capacity aggregation, transfer of licences and the extension of licensing to under 10-metre vessels in 1993 have also helped to reduce power but not tonnage. Ultimately the UK had only achieved about half of its MAGP objectives (EC Commission, 1996b). This chronic overcapacity, coupled with diminishing quotas, has created severe problems for the industry.

The cost of the 1988 PME soon began to cripple French investors. The authorities moved to adapt the PME and decreed that it was no longer necessary to withdraw old capacity to apply for a PME. A pool of available kilowatts would be shared between the regions each year and regional commissions would examine individual applications for permits. From 1993, ownership of old capacity no longer ensured a permit (Decree of 3 July 1993). Financial difficulties drove many out in 1992 and 1993 thus reducing capacity. However, France set up a decommissioning scheme in December 1993. She had not completely met her 1996 objectives but was expecting to achieve them by the spring of 1997.

After three fleet guidance programmes, both countries now have licensing systems. Decommissioning has been costly and, for a large part, ineffective in eliminating real capacity. In terms of policy-making in both countries, structural decisions were beyond the control and influence of the industry.

QUOTA MANAGEMENT

Member states are responsible for the management of their national quotas. The UK and France have adopted contrasting approaches. When the 1983 CFP regime was created, the UK already had the necessary legislative framework and some experience of quota allocation and management. A licensing scheme was put into place to regulate landings of pressure stocks in 1984. Quota management has gradually and increasingly involved POs in collaboration with the various fisheries departments and the local inspectors. Nineteen POs share the task of managing over 95 per cent of all UK quotas (Goodlad, 1996). Quota allocation operates on the basis of individual track records. Some observers see this development as a move towards greater industry involvement in fisheries management. However, this shift is not exclusive to fishing but forms part of the general trend under the successive governments since 1979 to involve private interests at the expense of the public sector (Willis and Grant, 1987). The UK has developed a comprehensive monitoring framework and has built up a very good record on compiling landing data and liaising with Brussels. However, as pressure on stocks has grown and quotas have been reduced, illegal landings have increased. They were estimated to represent some 40 per cent of all landings in 1996 (SCDI, 1996: 3).

A number of factors militated, in France, against the setting up of a quota management framework. Generous quotas, a traditional reluctance to use quotas as a management tool (Hamon, 1988), confusion and conflict in the respective roles of POs and CLPMs and industry disagreements resulted in policy inertia. Pressure from Brussels (EC Commission, 1986a), and increasing inter-user conflict, eventually forced the authorities to act in the late 1980s (Thom, 1993). A National Committee for Quota Management (now Fleet and Quotas Commission) was set up but only started meeting – informally – in 1988. As with the PME, this meeting was considered a cultural revolution for the industry. Since 1991, quota allocation has been formalized in an *arrêté* published every February. The distribution keys vary from stock to stock, but the general pattern is for an initial geographical division among five regional areas, then among POs and ports. Corporate track records are used, with any unallocated quotas remaining in a national pool. This situation reflects the common will to retain quotas as 'state property' (*Le Marin*, 17 February 1995). Due to disagreements between industrial and artisanal branches, and between both and the CLPMs, French POs have, hitherto, refused to manage quotas. Only half of French vessels, accounting for two-thirds of the value of all landings, are PO members. Overall, the management process remains conflictual and unsatisfactory, with a sizeable level of unrecorded landings. As the EU Commission underlined recently, rigorous monitoring of landings is still not a priority in France, a problem only mitigated by the underspent quotas in several stocks (Commission, 1996a).

CONTROL OF FISHING ACTIVITIES

Member states are responsible for enforcement of fisheries measures in waters under their jurisdiction. Proper enforcement requires a network of services from surveillance and inspection to prosecution. It needs resources and, above all, political will. The cultures regarding enforcement vary substantially in the two countries. In the UK, the determination to enforce regulations is illustrated in the resources committed to control. This task is shared between the fisheries departments, with each having defined areas to control. The UK had availed itself of the legislative framework to enforce measures well before the advent of the 1982 EC control regulation. There is no doubt that, for the EU Commission, the UK has the best record in this field and that the country 'provides an example of how the CFP should be enforced' (EC Commission, 1996a: 112).

Paternalism and a desire to protect fishermen have traditionally guided French fisheries protection services. Protection had to take precedence over sanction (Thom, 1993: 301). Attitudes may be changing, but slowly. Successive EC Commission reports have deplored the lack of progress in enforcement (EC Commission, 1986a, 1992, 1996a). There is no specific fisheries protection service in France. Instead, a great number of agents, services and government departments have been charged with some enforcement tasks (CCE, 1992). Fishing is seen as but one of the many activities in an area over which the French state exercises its sovereignty (Jégouzo, 1988: 375). This situation makes it difficult to evaluate the level of resources allocated to control. Anecdote and reports indicate a considerable degree of unrecorded and undersized fish landings, a poor return of log books, landing declarations and sale notes, poorly enforced technical measures and patchy transmission of monitoring information to the Commission. A comparison of inspection records shows the different priorities of French and UK officers. Thus, the UK determination to enforce quota restrictions is reflected in the bulk of registered infringements relating to log book and landing declaration offences. These accounted for about half of all UK vessel infringements in the early 1990s, leaping to 86 per cent in 1994. In France, this offence represented a mere 3–5 per cent of infringements by French vessels over the same period. Yet, half of detected infringements by French vessels inspected in UK waters related to this same offence (Thom, 1993: 305).

Assessing the effectiveness of enforcement is a notoriously difficult task (Sutinen and Hennessey, 1986). The UK record in this field is puzzling. Its commitment to rule enforcement is undeniable. Yet, this has to be contrasted with its disregard for its MAGP objectives and dealing with overcapacity. Illegal landings, log book and landing declaration fraud are often claimed to result from fleet overcapacity. In France, on the other hand, action on respecting MAGP targets has not been matched in enforcing rules in other spheres. Strangely, the outcomes are similar with high levels of rule breaking.

CONTRASTING PATTERNS OF INTERACTION

The relationship pattern between the UK government and the fishing sector does not appear to have changed since the CFP inception. The nature of the relationship has, however, deteriorated quite badly over the period. The nadir in the policy communities must have been reached in 1993 and 1994 during the government's attempts at imposing effort control as a main management tool. Various schemes were imposed on sections of the fleet before the Sea Fish (Conservation) Bill fiasco which had to be abandoned. The correspondence between the departments and the industry illustrates the gulf between the two sets of partners. The government was always prepared to hear the industry on the modalities of implementation but not on the *substance* of policy. Pressures led to divisions in the industry and the creation of new structures further weakening the position of the sector. The federations were split on crucial issues such as the UK withdrawal from the CFP and use of effort control as a management tool.

France paid dearly for her inertia in social and economic terms. Rioting, strikes and demonstrations shook the fishing sector in the winter of 1993 and 1994. Over-investment, the PME, currency devaluations and increased third-country fish imports led hundreds of vessels to the brink of bankruptcy. Fishermen's anger was directed not only at Brussels and Paris but also at their own representatives. Grass-root committees, tellingly named survival committees, were created. The French authorities responded in their usual fashion by making aid available, convening meetings, creating commissions to vet applications for aid and by reforming various structures in the artisanal sector. A new guidance law made its way through the French Parliament over the winter of 1996–7 to define the fishing policy environment. The rhetoric is reassuring, reaffirming the state's commitment to its industry. However, the reality is stark. The industry has been left divided and battered and, as in the UK, the future looks bleak.

UK and French fishing policies: a comparison

Differences in the policy-making processes of the two countries and in their traditional approaches to their fishing sectors are obvious. These have led to contrasting patterns of relationships in the respective policy communities. Both have chosen different paths to implement regulations, yet policies and outcomes have proved strikingly similar. Despite a pluralist mode of interest representation in the UK and a corporatist structure in France, policies were ultimately devised in a similarly authoritarian manner in both countries. Both industries spent their time reacting to proposals or fighting against legislation and their involvement was confined to the level of details. The *substance* of policies was not for discussion. There are further similarities.

Neither country has yet devised a fishing policy and implementation of CFP measures has been *ad hoc* and guided by self-interest. The transnational dimension consistently offered national authorities the justification they needed to force through unwanted measures. In the UK, market ideology, criticisms of the decommissioning scheme and the humiliation of having to suspend UK legislation because of EC law, all combined to fuel the government's hostility to Europe, the CFP and MAGPs. It is difficult to believe that the UK's record in quota management is the result of a normative desire to implement the CFP conservation regulation. Had this been the case, tackling overcapacity would have been made a priority, especially as illegal landings became widespread. It has often been argued that quota enforcement was part of a drive to force uneconomic enterprises out: in short, decommissioning without involving the Treasury. However, control is a costly exercise and there is a desire to ensure justice among producers. In this context, quota management was congenial to the UK authorities.

France chose not to implement CFP regulations until she had her back to the wall. There have, undeniably, been a series of cultural shocks for the authorities and industry alike. The creation of the PME in 1988 made the French realize that they could no longer ignore the CFP dimension. However, it had been principally the freezing of EC aid and the realization that overcapacity would soon threaten the economic performance of French vessels which had spurred the authorities into action. The EC dimension provided the justification for the measures adopted.

Policy outcomes were strikingly similar in both countries. Decommissioning funded the withdrawal of old, uneconomic units, with a negligible impact on the resource. These vessels would have left naturally, without any financial incentive. EU and national funds were, of course, being spent to destroy capacity that had been created ten years previously using funds from the same sources. The industry paid directly for the effects of such a policy. In the UK, operators acknowledged publicly, sometimes in court, that they had to break the rules in order to survive (*Fishing News*, 13 December 1996). Hundreds of their French counterparts had to go begging for a share of the aid programmes involving tax reductions and longer bank loan repayment times.

There are some areas, however, where differences in the nature of the licensing schemes and the allocation of quotas emerge. Licensing schemes have now been well established in both countries. Yet, they are very different in that market forces determine who gets UK licences, while, in France, legally, the authorities in consultation with the industry, allocate new licences according to a number of set criteria. There also remains a difference in the ownership of quotas which remain the 'property of the state in France', while licences and track records are being traded in the UK. Will these differences ultimately result in different outcomes? It is too early to say, especially after the UK fisheries minister stated that 'no-one has ownership rights over UK quotas' (*Fishing News*, 27 December 1996).

Two areas of EU legislation may make different measures result in similar outcomes. The first is that of flagships – vessels registered in one member state but with their economic and operational ties in another. Thus overcapacity in one country can be offloaded onto another. These vessels also receive quotas from their flag country. Already well implanted in the UK, the flagship phenomenon is currently well under way in France. The second area is the advent of the EU Special Fishing Permits, already necessary in some Western waters and for some stocks, and whose use is likely to be extended in 2003. What impact will they have in terms of quota allocation and ownership?

CONCLUSION

It appears that the sectoral policy approach which predicts that policies and politics show 'convergence across nations within sectors' (Freeman, 1985: 486) is vindicated in the case of fisheries management. Policies and outcomes are determined by sectoral imperatives. The nature of the resource itself and the degree of externalities on states make international cooperation necessary. Institutional arrangements, in this case common management in the EU, also matter, as do trade agreements. Despite the poor implementation record in the two countries, it cannot be denied that EU measures had an impact on policies and outcomes. The absence of political and economic saliency of the catching sector partly explains the neglect in the various arenas. Fishing is often used as a trade-off for loftier objectives. This aspect, coupled with the divisions in the industry, strengthen the authorities' hand. Fishing has an important role to play. However, no individual state acting alone will be able to ensure sustainable fishing for the future of coastal communities. Nothing less than a global approach to the exploitation and marketing of the resource has any chance of bringing fisheries management failure to an end.

NOTES

1. The terms 'artisanal' and' industrial' relate to two patterns of vessel ownership. Industrial vessels are company owned and crews receive a minimum wage to which a percentage of takings is added. In the artisanal sector, vessels are generally owned by individual skippers, families or partnerships. Crews are paid on a share basis.
2. Union des Armateurs à la Pêche de France, which is the industrial sector union.
3. Cooperation is the structure representing the artisanal sector.
4. Permis de mise en exploitation (PME).

BIBLIOGRAPHY

Atkinson, M. and Coleman, W. D. (1989) 'Strong States and Weak States: Sectoral Policy Networks in Advanced Capitalist Economies', *British Journal of Political Science*, 19: 47–65.

CCE (Comité Central d'Enquête sur le Coût et le Rendement des Services Publics) (1992) *La Coordination de l'Etat en Mer*, Cours des comptes, Paris.

Cmnd 1266 (1961) *Inquiry into the Fishing Industry – Report of the Committee on Fisheries* (London: HMSO).

Durand, J. L. et al. (1992) *Le Plan Mellick; Résultats au 17.02.1992 et Commentaires*, (Ifremer).

EC Commission (1986a) *Report by the Commission to the Council on the Enforcement of the Common Fisheries Policy* (COM(86) 301 final, 9/6/1986).

—— (1986b) *La Politique Structurelle dans le Secteur de la Pêche et de l'Aquaculture, Document de Travail* (SEC (86) 975 final, 12/6/1986).

—— (1992) *Report on Monitoring Implementation of the Common Fisheries Policy* (SEC (92) 394 final, 6/3/1992).

—— (1996a) *Monitoring the Common Fisheries Policy, Commission Report* (COM(96) 100 final, 18/03/1996).

—— (1996b) *Report on the Progress of the Multi-Annual Guidance Programmes for the Fishing Fleets at the End of 1995* (COM(96) 305 final, 01/07/1996).

Freeman, G. P. (1985) 'National Styles and Policy Sectors: Explaining Structured Variations', *Journal of Public Policy*, 5, 4: 467–96.

Goodlad, J. (1996) 'Sectoral Quota Management: Fisheries Management by Fish Producer Organizations' (this volume).

Hamon, J. Y. (1988) *Organisation Européenne en Matière de Produit à la Mer* in *La Mer: Hommes, Richesses, Enjeux*, Vol. II (Ifremer, ENA, Navfco (France)), pp. 6–79.

Hayward, J. E. S. (1982) 'Mobilising Private Interests in the Service of Public Ambitions', in J. J. Richardson (ed), *Policy Styles in Western Europe* (London: Allen & Unwin), pp. 111–40.

Hennequin, J. C. (1989) *Rapport sur la Modernization de l'Organisation Interprofessionnelle des Pêches Maritimes et des Cultures Marines* (CCPM).

House of Commons (1989) *Merchant Shipping*, Vol. 158, No. 161; 25 October 1989, pp. 993–1015.

—— (1992) *Sea Fish (Conservation) Bill*, Vol. 209, No. 23; 8 June 1992, pp. 40–119.

Jagot, L. (1996) Private communication.

Jégouzo, Y. (1988) *Les Rôles Respectifs de l'Etat et des Collectivités Locales en Mer in La Mer: Hommes, Richesses, Enjeux*, Vol. II (Ifremer, ENA, Navfco (France)), pp. 372–418.

Jobert, B. and Muller, P. (1987) *L'Etat en Action, Politiques Publiques et Corporatismes*, Presses Universitaires de France (PUF) Paris.

Jordan, A. G. (1990) 'Sub-governments, Policy Communities and Policy Networks. Refilling the Old Bottles?', *Journal of Theoretical Politics*, 23: 319–338.

—— and Richardson, J. J. (1982) 'The British Policy Style or the Logic of Negotiation', in J. J. Richardson (ed.), *Policy Styles in Western Europe* (London: Allen & Unwin), pp. 80–110.

Marsh, D. (1995) *State Theory and the Policy Network Model*, Strathclyde Papers on Government and Politics, No. 102.

—— and Rhodes, R. A. W. (1992) 'Policy Communities and Issue Networks – Beyond Typology', in Marsh, D. and Rhodes, R. A. W. (eds), *Policy Networks in British Government* (Oxford: Clarendon Press).

Meuriot, E. (1986) *La Flotte de Pêche Française de 1945 à 1983, Politiques et Réalités* (Ifremer, France).

National Audit Office (1987) *Financial Support for the Fishing Industry in Great Britain* (London: HMSO, 22 July 1987).

PAC (Public Accounts Committee) (1988) *Financial Support for the Fishing Industry in Great Britain* (House of Commons, 26 January 1988 (266)).

Rabot, J. (1993) Interview, Paris, 12 March.

Rodgers, P. E. (1989) *Comparative Implementation of MAGPs under EC Rule 4028/86*, EAFE, Annual Meeting, Lisbon, pp. 42–3, 101–2.

SCDI (Scottish Council Development and Industry) (1996) *The CFP Review Group Report* (Highlands and Islands Area Committee).

Scottish Fishermen's Federation (1983) *Response to Fisheries Departments' Consultative Paper on Restructuring and Fisheries Management of 30 March 1983* (25 May 1983).

Shackleton, M. (1986) *The Politics of Fishing in Britain and France* (Aldershot: Gower).

Sutinen, J. G. and Hennessey, T. M. (1986) 'Enforcement: The Neglected Element in Fishery Management', in Miles, E., Pealy, R. and Stockes, R. (eds), *Natural Resources Economics and Applications* (Washington: University of Washington Press), pp. 185–213.

Thom, M. (1993) *The Governance of a Common in the European Community: The Common Fisheries Policy* (PhD Thesis, University of Strathclyde).

Weber, J. and Antona, M. (1990) *French Fisheries Regulation Schemes in the EC Context*, IIFET, Fifth International Conference, 3–6 December 1990, Santiago.

Willis, D. and Grant, W. (1987) 'The UK: Still a Company State?', in Van Schendeln, M. and Jackson, R. (eds), *The Politicization of Business in Western Europe* (London: Croom Helm).

5 Can Quotas Save Stocks?
Tim Oliver

INTRODUCTION

Managing fisheries by limiting the amount of fish fishermen can take out of the sea seems at first sight a perfectly rational procedure. Catching too much fish damages stocks and their capacity to reproduce, so restricting catches is the obvious answer. Simply set a total allowable catch (TAC) for each stock, and when that is caught call a halt. But when one considers what a political and bureaucratic quagmire has to be created to implement and enforce such a policy, and at what enormous cost, its appeal is sharply reduced. I propose to examine the way the system works within the Common Fisheries Policy (CFP) and in particular within the UK, although some countries outside the EU have more successful experiences with TAC/quota systems than is the case in Europe.

TAC/QUOTA SYSTEM

The system of TACs and quotas within the CFP was set up primarily as a means of allocating resources and fishing rights to the member states. As Mike Holden (1994: 41) explains: 'The origins of this debate lay in the question of access; if exclusive fishery zones could not be achieved then a second-best arrangement would be to have guaranteed fishing possibilities.' This resulted in percentages of the annual TAC of each stock being guaranteed to each member state under a system known as relative stability.

The percentages were decided primarily on the basis of historic catches by the member states of the different stocks during a specified five-year 'reference period' (1973–8). This in itself gave a sign of what was to come in future years, as the member states argued among themselves for a reference period which would give the maximum historic catch records. The Hague Preference also gave another dimension to the allocations on the basis of 'vital needs'.

A further drawback of the uneasy compromise which was the 1983 CFP agreement was that the TAC/quota system was based on existing rates of fishing mortality, not their reduction. It was thus very much primarily a system of resource allocation, and only secondarily of conservation. Yet it became the basis of conservation in EU waters (with technical measures), with the tragic results we see today.

A system of enormous complexity but great rigidity was thus created as the means of controlling what is European man's last great hunting activity. Fish stocks fluctuate and shift with the seasons, the weather, tides and currents, water temperatures and salinity, feed and many other subtle and complex natural factors. Fish predate on one another and are thus highly interdependent. And it is rightly said that fish do not respect national boundaries. Yet we have divided up European waters into neat rectangles for the purpose of administering the TAC/quota system.

The rigidity of the system cannot be tampered with because of the dangers of upsetting relative stability – the allocation of fishing rights based on historic fishing patterns. But much of the fisherman's skill lies in reacting to this ever-changing ebb and flow in the balance of the various stocks and their movement, changing his fishing patterns, species hunted and methods used as the circumstances dictate. So we see fishermen finding plentiful stocks of some species yet being unable to land them (NB not *catch* them!) because they have no quota. Saithe in the northern North Sea and cod in the English Channel are two prime examples.

An absolutely fundamental flaw in the system is that it attempts to allocate fishing rights which are fundamental to fishermen's livelihoods on the basis of a theoretically precise science which is in practice anything but precise. Consequently the state of the stocks as experienced by fishermen on the fishing grounds is time and again out of step with the state of the stocks as indicated by the scientists via the TACs/quotas. This leads to mistrust on the part of the fishermen, and eventually to complete alienation from the system, as has happened among UK and indeed European fishermen as a whole.

I do not say this as a criticism of the scientists, who work within the limits of their resources and considerable knowledge and expertise to provide the best assessments they can of the stocks. Nor am I saying that quotas will not work under any circumstances: for example, I believe they may have a place in pelagic fisheries, which are usually carried out as directed fisheries on a single species. (But even here they must be based on accurate stock assessments and must be effectively policed and enforced.)

But in the context of Europe's complex, multi-species fisheries, quotas are, frankly, a nonsense. They force fishermen to either dump good fish for which they have exhausted their quota when caught with species for which they still have quota, or to save only the best fish and discard the rest if fishing is good. Or, as more often happens – because it goes against a fisherman's every instinct to dump good fish he has worked so hard to catch – it is landed illegally, and this in turn means that the landings data on which many of the scientists' assessments are made is deeply flawed, leading to even less accurate assessments and so on in a downward spiral.

ALLOCATION OF QUOTAS

Another major problem with quota systems arises in the method of allocating them to fishermen. In the UK this is done through the 'track record' system, which relies on a fisherman's 'historic' catch record over a given period – currently a 'rolling' three-year period (two years for pelagic and distant-water vessels). Basically this means that the more a fisherman catches the more quota he is entitled to – hardly a formula for encouraging moderation. Quotas become targets to be achieved rather than ceilings to limit catches. Fishermen dare not undershoot their quota allocation, as this means they will get less quota next year – much as financial managers always spend their budgets for fear of finding they have been reduced in the next round.

This adds more inflexibility to the system and actually increases the pressure on pressure stocks. If a good fishery develops on a non-quota species – for example, the deep-water fisheries to the west of Scotland – or a fishery less under pressure when a seasonal fishery is under way for example, the fisherman cannot leave the fishery where he has a limited quota until he has caught that quota because he may not get another chance. So fishing remains intense on the pressure stock, while the more abundant stock is under-exploited.

A track record system also causes problems when a fisherman has to lay up his boat or cannot fish it for a period, for whatever reason, as his catch record will then diminish, leading to a smaller allocation next year. Small vessels which cannot range far from home and which are limited to seasonal inshore fisheries are particularly vulnerable under the track record system, as a poor fishery or prolonged bad weather can severely curtail catches with the same result.

The track record method has one further disadvantage: while fishermen regularly under-declare their catches to stay within their quota 'on paper', many over-declare their real catch in order to build up a track record 'on paper' which is in excess of their actual catch. This practice has become known as 'ghost fishing', and it leads to unfairness and a further distortion of the actual catch figures with which the scientists have to work. The problem has intensified dramatically in the last few years or so because of the growing value of quotas and track records which are now attached to vessel licences and in which there is in the UK a healthy trade. The UK currently has an informal quota transfer system which is in effect an embryonic individual transferable quota (ITQ) system for over 10-metre vessels in membership of a fish producer organization (FPO), although unacknowledged as such and not administered through a formal mechanism. (I shall return to the subject of ITQs later.)

This informal ITQ system is in turn affecting the structure of the fleet and its ownership. Licences are finite and all types are thus becoming ever more expensive. Thus only the most successful fishermen in the biggest boats can afford them, or – as the industry frequently complains – the flagship operators. This is inhibiting the renewal of the 'bread and butter' sector of the fleet, the 40–60 ft prawn and white-fish trawlers, because owners almost always need to aggregate extra licences when they build new boats, wanting more power or a bigger boat, and they are priced out of the reach of most.

QUOTA MANAGEMENT

Quotas in the UK are currently managed under a system of 'sectoral quotas' whereby each year's UK allocation of each species in each ICES area is shared between 19 FPOs, the remaining 'non-sector' boats over 10 metres and vessels under 10 m.

Each FPO's allocation is based on the track record of its individual member vessels. FPOs manage their quota allocation as they see fit: some give each boat its individual quota on one or more species to take through the year as its skipper wishes, and some allocate monthly amounts which are adjusted according to uptake, season, etc. This can be the source of much friction, because fishermen in the same port and working the same fishery could find themselves with widely differing quotas depending upon how their FPOs had managed their quotas. But less argument is heard nowadays about this issue, since quotas have reduced and as the 'black fish' trade has become more established.

If an FPO overshoots its quota for any species in any area it is penalized. FPOs can swap quotas (using a 'cod equivalent' basis) with each other and with the non-sector. If an FPO does not take up all of its quota, fisheries departments can reallocate the unused quota on a pro-rata basis to other groups. If a vessel leaves an FPO and joins another – usually because of a change of ownership – it takes its track record with it, based on its catch up to the year end. FPOs can now 'ring fence' quotas, that is purchase a member vessel's licence (which has that vessel's track record attached) and retain that vessel's quota within the FPO to be fished by the FPO members as a whole. The large number of vessel movements in the fleet currently taking place and the associated licence and quota track record movements obviously add considerably to the complexities of the statistical work involved in administering the quotas.

There is a growing belief among the FPOs, however, that the track record system should be abandoned in favour of fixed allocations because of the negative effects of the track record system. Such a system was recommended by the CFP Review Group which reported in July 1996, and recent discussions between the industry and ministry officials have resulted in a consensus that the current system will have to change. But any change is unlikely to be in place before 1 January 1998.

The non-sector is managed by the fisheries departments, in conjunction with various fishermen's organizations. Quotas are allocated on a monthly basis, and adjusted according to circumstances. One important difference between the non-sector and the FPOs is that non-sector fishermen can be prosecuted by fisheries departments and tried and fined in the courts for landing over-quota fish, since they have exceeded government limits and committed a criminal offence. By contrast, FPO members are subject only to the discipline of their FPOs if they are caught landing over-quota fish, although they can be prosecuted by fisheries departments for such offences as incorrect logbook or landings declarations. Some FPOs have begun to fine members who are caught landing over-quota fish, on occasion quite heavily, but in general it appears that a fairly benign regime exists, with a blind eye being turned most of the time. And even if sector fishermen are fined, they do not acquire a criminal record for landing over-quota fish, unlike their non-sector colleagues.

The under 10 metre sector has a separate allocation, but there are no monthly limits set on catches as in the over 10 m non-sector because of the large number of vessels concerned. The fisheries departments either close the fisheries occasionally to spread them over the year and ensure quota is left for seasonal fisheries, or close a fishery when its quota is exhausted.

Both the over and under 10 metre non-sectors have suffered a depletion of their quotas over the years, as bigger boats with greater range have acquired track records and then joined FPOs, taking the accumulated quota track record with them. Particular fisheries which have suffered from this have been the North Sea cod and sole fisheries, the Channel sole fisheries and the SW mackerel handline fishery. North Sea sole has been a particularly sore point, as most of the North Sea quota was taken by Dutch-owned beam trawlers flagged on to the UK register, leaving the small boat non-sector fishermen with farcical monthly quotas of as little as 250 kg.

Non-sector fisheries also suffered in the past from misreporting of catches by FPO vessels of species for which the FPO did not have a sectoral allocation. This has been rectified by fisheries departments requiring FPOs to take sectoral allocations of all pressure stock species. The non-sector is now protected by having certain quotas 'underpinned', i.e. a basic level set below which quotas cannot fall, regardless of TAC fluctuations.

All this results in a very complicated statistical exercise, both in the allocation of the figures and the monitoring of the uptake each year. It is quite normal for fish producer organizations not to know their final allocations until April, May, June or even later, and have to manage their resources on 'guesstimates' in the meantime.

There is another complication: the distinction between analytical and precautionary TACs. There are just over 100 separate TACs allocated in European waters, yet fewer than half of these TACs are fixed on the basis of scientific analysis of the stocks in question each year (analytical TACs). To provide an accurate assessment of all these stocks every year is a tall order, so in the case of 56 TACs – more than half the total – they are set on a precautionary basis, i.e. without specific scientific analysis but on the basis of the best available knowledge. Precautionary TACs are in fact set primarily for political not conservation reasons: without a TAC the stock cannot be allocated to the member states, so fishermen from any state could fish anywhere for them and thus relative stability would be undermined. They do have an impact on conservation, however, in that fishermen catching non-TAC species would have to discard those TAC species they caught for which they did not have a quota, thus undermining the analytical TACs.

Under the CFP, any fish which fishermen catch for which they do not have a quota must be discarded; it is not illegal to *catch* such fish, only to *land* them. But it is landings data on which the TACs are based, not catch data, so there is an immediate distortion in the figures. Some allowance is made in the calculations for discards, but again this is only a rough estimate.

Clearly, then, the management of the TAC quota rests on a dubious basis. Yet upon this base is built a precise statistical system which eventually allocates to some UK fishermen quotas for some species measured in amounts as small as 250 kg per month.

There is nothing wrong in principle with attempts to assess stocks in the course of fisheries management. What is wrong is that the entire system of EU fisheries management is based on such an enormously complex, crude, cumbersome and inflexible system. It is on 'the scientific evidence' that the livelihoods of thousands of fishermen and many more thousands of ancillary workers depend.

What is inevitably a rough-and-ready mechanism is used to 'fine tune' catches, and is treated by fisheries managers and politicians as though the data were the result of experiments undertaken in controlled laboratory conditions.

THE POLITICIZATION OF THE CFP

One reason why the TAC/quota has failed is because of the powerful political element in fisheries management. The creation of so-called 'paper fish' by politicians, over and above the scientific recommendations, in the early years after the 1983 CFP settlement accelerated the erosion of the stocks. The politicians were reluctant to accept scientific advice to reduce catches for fear of courting political unpopularity at home. Politicans take a short-term view because they have no vested interest in taking a long-term view. They are unwilling to accept unpopularity by taking tough decisions only to let others reap the benefits in the future.

This point has been made very strongly in a recent paper by Ad Corten, a scientist at RIVO, the Netherlands Institute of Fisheries Research. Mr Corten is an experienced scientist who, among other roles, has been on the ICES Advisory Committee on Fisheries Management, which advises the European Commission on TACs. Corten notes that because fisheries managers avoid difficult decisions, fisheries scientists are disillusioned with the management system and have lost interest in it, inevitably leading to even poorer data and results. Many scientists increasingly prefer to 'devote their time and energy to interesting scientific problems, rather than to biological book-keeping of fish stocks' (Corten, 1996: 14).

The gradual erosion of the stocks throughout the 1980s was masked to a great extent by the rapid growth in fishing technology and power in this period – aided it must be said by an easy availability of grants for modernizing vessels and building new vessels. Catch rates were thus maintained, creating a degree of complacency.

But political attitudes have hardened in recent years as the weakening stock situation has become more transparent, and also because of the growing political pressure from the environmental movement on European and national politicians to address marine conservation. This has led politicians to heed more closely the scientific advice and be much less willing to appease fishermen. Consequently most quotas have continually reduced, but fishermen have been unwilling or indeed unable to adhere to them. They have thus 'cooked the books' (logbook landing declarations) and distorted the data on which the scientists rely. The scientists have in the case of the main west of Scotland demersal stocks refused to make TAC recommendations and simply said there must be a 30 per cent reduction in fishing mortality on these stocks.

THE PSYCHOLOGY OF FISHERMEN

One factor in my view which has always been totally neglected by the bureaucrats and politicians who run our current quota regime is the psychology of fishermen. Fishermen go to sea to catch fish: that is the whole point of their activity. They do not invest large, often vast and mostly borrowed sums of money, and endure the hardships involved in being a fisherman, to catch fish and then throw them away. When fishermen are faced with quotas which they consider to be inadequate, either because they find more fish in the sea than the scientists tell them is there, or because they simply have to land more to stay viable, they will get round the quotas. That is a fact of life. It is often said that quotas create waste by forcing fishermen to dump fish, and a certain amount of dumping does occur. (Saithe in the northern North Sea is a current notable example.) But it goes against the grain for a fisherman to throw away good, marketable fish simply because he is out of quota for it. He has caught it anyway, and it will be dead or will die when he throws it back, so what is the point?

No, the fisherman who catches fish for which he has no quota will land it 'through the back door', and there are well-developed mechanisms for doing this, to the detriment of the scientific data and orderly marketing.

The demersal fisheries in EU waters are very largely mixed fisheries, so it is virtually impossible for a fisherman to catch only those species for which he has a quota and not catch others for which his quota is exhausted. Elaborate and complicated by-catch rules attempt to get round this problem in some fisheries, but the reality is that the majority of what is caught of a marketable size and quality is landed and sold. A fisherman will either simply not declare at all an over-quota catch of a given species, or will report it as another species for which he does have quota or which is a non-quota stock. Another way round the system is to report fish as having been caught in an area where a fisherman does have a quota, when in fact it has been caught in an area where he does not.

ENFORCEMENT PROBLEMS

Stricter enforcement is the answer to this problem say the supporters of quota systems, but this is more easily said than done, particularly in countries like the UK and Ireland which have enormous coastlines and a multitude of ports where illegal fish can be covertly landed. Enforcement costs money, and there are clearly restraints on how much money any government can or will allocate to fisheries enforcement when it has so many other demands on its resources. Fishery officers in the UK are not highly paid, and many have become alienated from the system they are supposed to enforce. Ports cannot be manned round the clock and fishermen can usually land their over-quota fish without too much difficulty – although it has to be said that they hate having to operate like this.

The quota system is complex, and the fish trade is a diverse and fragmented business in which fish is often landed in one port and sold in another, either at auction or directly to processors. This makes the job of tracing illegal fish difficult, and fishery officers often have to undertake many weeks or even months of patient 'detective work' before a successful prosecution can be brought. All this costs money, and still does little to control the situation. Again, penalties could be increased to act as a greater deterrent, but this raises ethical questions. Maximum penalties of £50 000 are already available to magistrates for many over-fishing offences, although fines are usually nowhere near this level except in the case of serious and persistent offenders.

Licences could be withdrawn, but that means taking away a fisherman's livelihood. Are such harsh penalties justified against men who in the end are trying to make a living by supplying food? How would they compare with the light penalties so often imposed on serious offenders – rapists, thugs, muggers, drug dealers, etc. – which arouse such public anger?

Continuous satellite monitoring of fishing vessel movements is seen by the EU Commission as the answer to the problem of vessels misreporting the area in which they have caught their fish, but this is fraught with problems and is meeting stiff resistance from the UK industry. And it must be asked whether a system which requires such costly and 'high tech' measures to make it work – albeit only partially – is a good system. A case of 'taking a sledge-hammer to crack a walnut'?

INDIVIDUAL TRANSFERABLE QUOTA (ITQ) SYSTEM

Would a full-blown ITQ system be the answer? I do not want to go too deeply into the detail of an ITQ system because other chapters cover this topic, particularly the issue of property rights in fisheries. But it does arouse very strong feelings among many fishermen, particularly north of the border, where ITQs are seen by most fishermen (not all) as a threat to their traditional freedom of the seas and way of life.

Fishermen fear that ITQs will result in their industry passing inexorably into the hands of a relatively few owners, resurrecting the company ownership of the industry which prevailed in the now almost defunct distant-water sector. Experience in countries like

New Zealand indicates that this would happen, and the current rapidly rising price of licences with their attached track records in this country confirms this indication. British fishermen also fear that much of the industry would pass into foreign hands, as there appears to be much more finance available in countries such as Spain to purchase quotas than is the case here. Again, current experience seems to bear this out. It is a common complaint in the industry that one of the reasons for the current high cost of licences is that flagship operators always seem able to bid up their prices beyond the reach of the ordinary skipper/owner.

No one is ever forced to sell his licence of course, but the rank-and-file fisherman, the traditional skipper/owner/part-owner who forms the backbone of the industry, has little effective choice. As prices rise, they become too tempting to resist, and with all the other pressures facing them many fishermen are tempted to sell up.

I think that many of the disadvantages of the current quota system I have outlined above would also apply to an ITQ system, although they would be ameliorated to some extent as the effects worked through the system. Such a system would still depend on scientifically assessed TACs and the inadequacies of these would still apply. However, they would become less critical as the fleet diminished and stocks in theory at any rate improved. But that would take some time and in the meantime there would be political, social and economic stresses as the fleet contracted and moved into foreign owership, and traditional fishing communities declined. ITQs would have to be administered and policed, still a costly exercise given the complexity of UK fisheries and area of our waters and coastline. Arrangements would have to be made for compensation if TACs were reduced and quotas for which owners had paid substantial sums could not be fished. Furthermore, all kinds of complex quota leasing and transfer arrangements are invariably involved when ITQ systems are set up, which would make the system even more complicated than the current arrangements.

ITQs offer a market solution but is a market solution what is required? The need to maintain the jobs and economic infrastructure of fishing communities in remote coastal regions where little or no alternative employment is available should surely also be a policy priority.

In a new chapter added to a reprint of the late Mike Holden's book on the CFP, David Garrod, the former Director of Research at MAFF's Lowestoft laboratory, doubts that an ITQ system could work within the multinational CFP system. 'If the broadly allocated TACs cannot be enforced at present, there is no prospect for doing so within an ITQ system. In a multi-national fishery, ITQs would be violated just as frequently as the current TAC, perhaps more' (Garrod, 1996: 273).

CONCLUSION

Despite all the problems with the current quota system and its failure to deliver, I cannot see it changing significantly before 2002 and the expiry of the 20-year 1983 CFP agreement. Even then it will be difficult to change, because by 2002 so many people will have such huge investments in quotas that the pressures to retain the status quo will be enormous. The CFP Review Group suggests that the problems could be overcome by allocation of quotas to FPOs only and not to individuals. This would still give fishermen access to the quotas they have brought into the FPO but would prevent individual ownership and also sales to flagship operators. However, with the likelihood that fixed quota allocations attached to vessels' licences will be introduced in the UK in the near future, it must be open to question whether owners with good quota allocations would be willing to pool them in their FPO with less fortunate members of the FPO. Many in the UK industry believe that fixed quota allocations, if introduced, will lead inexorably to a full-blown ITQ system, and that this, indeed, is the government's long-term aim.

From a conservation point of view, quotas will only work if they are high enough to allow fishermen to work within them. Once they become restrictive, fishermen find ways to beat them, hence the current weak state of European stocks. So if fishing effort has to be reduced, as it does now in EU waters, it follows that there would still have to be a hefty reduction in the size of the UK fleet for quotas allocated to FPOs – or allocated in any other way – to work as intended to conserve stocks.

Fishermen's alienation from the system is also increased by the unwillingness or inability of other member states to enforce the restrictions on their own fishermen. Spain and France are particular 'bogeymen' in this respect. Even though allowance must be made for the prejudices of British fishermen, it is clear that enforcement of regulations in these countries is poor, and that a different culture and attitude exists towards the enforcement of EU regulations in general – and not just in fisheries. This would be a problem whatever system of conservation was introduced, because if the political will to enforce the regime is absent or feeble, as it clearly is in some member states, then no regime will work. But quotas are difficult to enforce even when the will is there to do so – without it the task is hopeless.

However, because of the allocation function of TACs/quotas, changing the system radically will be extremely difficult. Politicians will not want to get involved in radical change, nor will officials, either nationally or in the European Commission, because it would effectively mean renegotiating relative stability, in fact a wholesale reconstruction of the CFP. Bearing in mind the eight-year marathon of negotiating the 1983 agreement, this is something which the fishery managers will avoid like the plague, and hope they will have moved on before the job has to be tackled.

Nevertheless, it seems inescapable that at some stage effective conservation policies will have to be directed towards limiting the effort that goes into the sea rather than trying to control the amount of fish that comes out of it – but that is another subject.

BIBLIOGRAPHY

CFPRG (1996) *A Review of the Common Fisheries Policy Prepared for UK Fisheries Ministers by the CFP Review Group*, 2 vols (London: MAFF).
Corten, A. (1996) 'The Widening Gap between Fisheries Biology and Fisheries Management in the European Union', *Fisheries Research*, 27: 1–15.
Garrod, D. (1996) 'The CFP Now', in Holden, M., *The Common Fisheries Policy: Origin, Evaluation and Future*, reissued with a new chapter by D. Garrod (Oxford: Fishing News Books).
Holden, M. (1994) *The Common Fisheries Policy: Origin, Evaluation and Future* (Oxford: Fishing News Books).

6 Individual Transferable Quotas and Property Rights[1]

R. O'Connor and B. McNamara

INTRODUCTION

For hundreds of years, there has been a public right to fishing in tidal waters. This was not always the case. According to Anthony Scott (1989), medieval kings and barons could grant fishery rights anywhere. However, if the fishing was too far offshore the right was difficult or impossible to enforce. As a result of this custom, many inshore and tidal fisheries existed, particularly in the eleventh to the thirteenth centuries, and the rights to them were being successfully enforced, just as in inland fisheries today.

Scott goes on to state that the development of exclusive rights of fishing in common law, at least for tidal waters, was 'snapped off' in 1215 when the Magna Carta was signed. In consequence there was a reversion to an earlier pre-1066 Saxon public right of fishing in tidal waters, a tradition that became entrenched in law and subsequently spread throughout British colonies and most of the world. Except in a band called the territorial sea, usually three miles wide, a fisherman from the adjacent coastal state had no better right to fish than one from a distant state.

There were some exceptions to the open access system. In areas such as coastal eastern and south eastern areas of Asia and the Pacific islands, where fish and other marine animals form the major source of animal protein, exploitation of fish has always been, and still is, strictly controlled by systems of local sea tenure (Ruddle, 1989). Under these systems, known as TURFS (territorial use and rights to fisheries), members are forced by custom to take into account the effect of their actions on community activities and returns. Special strictly enforced local rules apply to the allocation and conservation of supplies (Panayotou, 1989).

The open access system persisted over the centuries throughout

81

most of the world. It was favoured by the great naval powers who wanted freedom of the seas for various purposes. It was also a very desirable system for the advanced fishing nations. With large boats and what was regarded as an inexhaustible supply of fish, the high seas of the world became their oyster.

However, as technological developments improved and demand for fish increased, it came to be seen that fish stocks were far from inexhaustible. Recent advances in fish finding, navigation, propulsion, materials for nets and lines and in fish handling have been dramatic. In addition, the steady progress of naval architecture has made vessels safer, more efficient and longer ranging. These improvements plus a general increase in the real price of fish products have induced fishermen to create fishing power far beyond the capacity of fish stocks to sustain economic catch levels.[2] In the past three decades, many of the world's rich fisheries have been exploited to the point of near exhaustion. Such developments have strengthened the arguments of resource managers and fishery economists for a change in the open access rules (*The Economist*, 19 March 1994).

CONSERVATION DEVELOPMENTS

Faced with the massive over-exploitation of fishing resources, Governments and the European Union have in the past introduced a variety of policy measures for management of diminishing fish stocks. These measures fall into three groups:

- output controls;
- input controls; and
- technical measures.

Output controls constrain the catch of the fleet through the introduction of total allowable catches (TACs). Input controls limit the inputs used to produce the catch, such as the licensing of boats, restrictions on size and other dimensions of fishing vessels. Technical measures impinge on the relationship between inputs and output such as net size and type, area and time closures.

By and large, these measures have proved ineffective. In the management of stocks there has been a need to take account of the incomes of fishermen, many of whom belong to poor regional communities largely dependent on fishing. Managers have been unwilling to follow scientists' advice on TACs. Where countries have banned foreign vessels from their 200-mile exclusive zones, domestic fleets have expanded to take their place. Other countries have resorted to 'quota hopping' – registration of vessels by foreigners under flags of convenience in order to gain access to other countries' fish quotas – and it is proving almost impossible to counteract this development (McLysaght, 1991).

Governments have encouraged excess fishing by subsidizing fleets, often as a form of regional aid and in response to falling catches. Application of the EU proposals under the Common Fisheries Policy for reduction in gross registered fleet tonnage has had the perverse effect of increasing rather than reducing fishing power. Old, inefficient and sometimes tied-up vessels have been replaced by modern highly efficient craft fitted with all modern devices involving what has been described as 'capital stuffing' (see Townsend and Pooley, 1995). As the European Commission itself acknowledged in a Communication to the European Parliament on 6 December 1990, there is at present no effective control over fishing effort.

The failure of conservation policies is part of a more general phenomenon of over-exploitation of environmental resources attributable in large degree to shortcomings in the ways in which society and its members organize the use of those resources. Central to this is the way in which open access systems lead to human behaviour which, far from encouraging maintenance of sustainable resources, results in their systematic destruction – what has been described by Hardin (1968) as the 'tragedy of the commons'.[3]

A recent response to the 'tragedy of the commons' metaphor was to propose that public fish resources should be privatized through the introduction of vessel quotas, the argument being that only under such a system would the owners have incentives to protect the flow of service from resources into the future. Private ownership of quotas would protect the resource user from the open access 'prisoner's dilemma' trap, which results from disrespect for authority because of perceived past injustices and not knowing what other fishermen will do (see Kaitala and Munro, 1995). The early boat quotas were non-transferable but later in a number of countries these were made transferable, conferring property rights on the recipient.

/IDUAL TRANSFERABLE QUOTAS (ITQS) AND
JPERTY RIGHTS

According to Scott (1989), the catching rights conveyed by ITQs
are analogous to the territorial interest held by farmers and their
tenants. They form part of the owner's real estate.

Different writers distinguish a number of characteristics in real
property. One of these is transferability and that is why ITQs have
come to be defined as real property. A non-transferable boat quota
which could be changed or withdrawn at any time would not fall
into this category. However, some people would argue that the boat
quota is the important management tool. If this is so, why has transferability become such a regular feature of individual quota systems?

ADVANTAGES AND DISADVANTAGES OF TRANSFERABILITY

In their examination of the New Zealand ITQ system, Clark et al.
(1988) considered that three major objectives of management are
enhanced by the transferability of quotas. The first is to achieve
the optimum number and configuration of fishing vessels and fishing gear. A fisher will demand or offer to sell quota in whole or
part depending on whether the net revenue he expects to earn from
his last unit of catch is greater or less than the market price of the
quota. In this way, quota will gravitate towards the most efficient
operators and eliminate excessive vessels from the fleet.

The second objective of fishery management is to achieve a level
of catch which maximizes benefits to the nation while ensuring a
sustainable fishery. A fisher cannot control the quantity of fish to
be captured by his gear. Therefore, as he or she approaches the
limit of the individual quota holding, problems will arise involving
unintentional by-catch, quota busting and under-catching of quota.
Together these would amount to a major problem of deviation from
the TAC limit unless free pooling or transfer of quotas between
fishers were allowed.

The third objective of management is to achieve its goals at
minimum cost. In New Zealand, transferability has resulted in savings in the cost of both implementation and enforcement. It is difficult
to allocate fish quotas on the basis of historic catches. Mistakes
may be made but this problem should be self-correcting through

quota trading. In terms of the enforcement costs, the New Zealand authorities have introduced an audit scheme which makes it possible to check individual catches and save expensive surveillance at sea.

Transferability is also supposed to allow fishers to enter and exit the fishery more easily. A new fisher may enter the fishery by purchasing or leasing quota. Anyone retiring from fishing may sell or lease his quota.

The main disadvantages of transferability relate to social issues. Fishery management is not simply concerned with fish. It also has welfare implications for coastal communities. Speaking at a recent marine food conference in Ireland, David Thomson (1996) said that with the introduction of ITQs, fishermen are very pressured to sell their quotas and the result of this policy will simply mean the concentration of fishery wealth in fewer and fewer hands.

Most fishermen operate from small coastal villages for which they provide the core economic foundations. Hence the loss of local fishery fleets would have serious knock-on effects for small communities. This development would result in large welfare costs as well as in social dislocation. As a result of the blue-fin tuna ITQ programme, introduced in Australia in 1984, quota trading led to the exit of the New South Wales fishers by 1987 and to a reduction of over 50 per cent of the quota owned by Western Australian fishermen. This has led to a concentration of quota in South Australia, the share of the TAC increasing from 66 per cent in 1984 to 91 per cent in 1987 (Geen and Nayar, 1988).

The sale of quotas by the smaller fishermen is a serious problem, but high entry costs under an ITQ system pose even greater difficulties. Fishermen who are fortunate enough to get a permanent private share of a quota have a valuable asset which can be leased or sold. Elderly fishermen or those with other interests can sell rights they have not paid for, whereas a newcomer to fisheries will have to pay very high entry costs. In Iceland prices for the leasing of quota have risen enormously for all species. In 1992, leasing prices for cod and saithe equalled about half the landing prices of these species but in 1994/5 the leasing price of cod quota was 70–80 per cent of the average value of cod catches (Eythorsson, 1996). In a situation like this, potential entrants have no chance. Indeed, many Icelandic quota holders have given up fishing and lease their quotas to the large well established vertically integrated companies who may land their catches in other parts of the country.

In addition to the socio-economic factors mentioned above, there are a number of management problems associated with ITQs. Among these are:

- *quota busting* – the flouting of quotas through inadequate monitoring, enforcement and penalties and the misreporting of the areas where the fish are caught. As Rettig (1989) says, where the fish catch is valuable and is landed at many ports, enforcement difficulties may wreak havoc with individual quota programmes. Townsend and Pooley (1995) point out that under an ITQ system, the individual fisher has great incentive to avoid the restrictions on harvests by under-reporting landings;
- *high grading* – the discarding of low quality fish; and
- *by-catch problems* – the unavoidable catching of non-targeted species and their dumping at sea.

PROBLEMS WITH THE INTRODUCTION OF ITQs

There has been a general resistance in many countries to the introduction of ITQs. Reasons for this resistance include controversy over:

- the privatization of a traditional public resource;
- constitutional rights of government to transfer public rights to some individuals rather than others;
- the basis adopted for establishment of quotas;
- the distribution of property rights at the expense of those with traditional interests.

The question of how quotas are allocated is also very relevant. The initial allocation process is one of the most difficult, time-consuming and costly aspects of implementing an ITQ programme. This allocation is controversial because it determines who will receive the benefits of the programme, thus creating a valuable asset for some and excluding others. Initial allocations are usually made on the basis of catch history but such allocations raise the problem of how to measure and validate historical catches (Libecap, 1989). The result is that many participants appeal their allocations, thus delaying implementation.

A further issue which has arisen in Norway (Trondsen and Angell,

1992) is the political problem associated with ITQs. Assignment of property rights has important distributive effects and large-scale transfers of such rights will bring political opposition from vested interests created by the current regulatory scheme. Thus, the introduction of ITQs has been resisted in many countries by fishing crews, fish buyers and the large processors.

Some fishermen's unions claim that because crew members have traditionally shared in the catch, they should be entitled to part of their boat owner's quota. Countering this argument, Grafton (1996) says that assigning both crew and vessel rights would be difficult to implement as records of catches do not include crew members. Furthermore, monitoring quotas would be problematical if some crew members sold their quotas to other vessels.

Large processors object to ITQs because, under traditional systems, they could buy cheaply the big volumes of fish coming on the market in a short space of time. These volumes require large freezing and handling capacities and the financial ability to hold an inventory to sell during the rest of the year. Under an ITQ programme, a substantial number of smaller, less capitalized firms can enter the sector and switch from a predominantly frozen to a mainly fresh market, thus increasing price and reducing the market share of the large operators.

Despite these objections from the fishing industry, ITQs have been adopted in many fisheries throughout the world and, as pointed out by Scott (1989), the new systems, where they exist, have been steered into place, or at least welcomed, by public servants. Bureaucrats and politicians, overwhelmed with the difficulty of what they are asked to do, have open minds about new ways of doing it. In any case, politicians can often safely disregard fishery issues, since fishermen are minorities in most constituencies. Finally, national politics may be indifferent; a politician's governing party will probably be unconcerned about fisheries policy so that quotas and such like will usually be local issues.

INDIVIDUAL TRANSFERABLE QUOTAS AROUND THE WORLD

Various forms of individual quotas have been implemented for a wide variety of fish species in Australia, Canada, Denmark, Iceland, Ireland, the Netherlands, Norway, Portugal, New Zealand, the UK and the USA. Currently, there are over 50 individual quota programmes operating in fisheries around the world, each with different design characteristics depending on the circumstances. So far, however, only New Zealand and Iceland have put ITQs into practice as an overall national management system.

In the following pages, we describe the New Zealand and Icelandic systems. We also discuss two other programmes, one of which appears to be successful, the other less so, to show how individual quotas work at subnational levels. Some general views are then expressed on the working of ITQs generally.

The New Zealand ITQ system

Before the declaration of the 200-mile zone, the New Zealand fisheries were very small and confined to an inshore domestic industry up to 12 miles offshore. Beyond this distance the fisheries were exploited by foreign vessels from Japan, Korea and the Soviet Union (Clark et al., 1988). When the 200-mile zone was declared in 1978 the government was faced with developing management strategies for the fish resources of a very large and unfamiliar area.

Initially, a policy of limited domestic expansion, joint venture, arrangements and licensing of foreign fleets was applied to the zone outside 12 miles. In 1983 economic and biological problems in the inshore fishery reached such a critical stage that they could no longer be ignored. To cope with this and various other situations, the Fisheries Act of 1983 was passed. Under this Act the government introduced a management system for deep-water trawl fishermen based on individual company transferable quotas. This system later served as a model for the management of the stressed inshore species.

Initially the ITQs in the deep-water fishery were for a period of ten years and for seven key species. They were allocated on the basis of prior investment. In 1985, these quotas were made valid in perpetuity and other species in the inshore fishery were brought under ITQ management. The new allocations were calculated on the basis of historical catch, modified by the results of a buy-back scheme by the government, where it was necessary to adjust the TAC because of mismatches between fleet capacity and available catch. To facilitate quota transfer between fishers, a government-backed Quota Trading Exchange was established by the New Zealand Fishing Industry Board. However, most trade in quotas is done by companies negotiating directly with each other.

Quotas are managed on the basis of information contained in three documents – the catch landing log, the quota management report and the licensed fish receiver's return. The catch landing log provides an on-site record which must be completed by the skipper as soon as the catch is landed. The licensed fish receiver's return must be submitted to a registration office at least monthly. It shows the quota holder's name and the green weights of all fish received from each holder. A licence may be revoked if the licensee (fisherman, fish buyer or exporter) has been convicted of any fishery offence. The quota management report must be completed by the quota holder and submitted to a registration office at least monthly. It details by area the quantity of fish caught for each species for which quota is owned or leased. Since the different reports from quota holders, fish receivers and fishermen can be checked against each other, all play a part in monitoring the use of the resource. The monitoring programme is aided by the fact that over 85 per cent of New Zealand's fish harvest is exported.

Critics of the ITQ system stress the enormous difficulties entailed in accounting for the individual fisherman's landings. The New Zealand experience, however, demonstrates that a thorough auditing system involving log books, landing records, processor records and export licensing can provide the information necessary to obtain satisfactory compliance. The costs of enforcement are paid by the fishermen and the trade. In 1988/9 the government collected a little less than 10 per cent of the landed value of the fish caught by the New Zealand fleet. This, along with foreign access fees and other charges, was sufficient to cover the total operating budget of the regulator (MacGillivray, 1990).

In a comment on the New Zealand system, Huppert (1989) says that while the paper audit system may be satisfactory for New Zealand, it does not satisfactorily monitor at-sea dumping of small fish (high grading), at-sea processing operations aboard domestic or foreign ships and 'over-the-side' transfer of unprocessed fish at sea, particularly to foreign boats. It remains necessary to employ the 'game warden' approach to these operations.

However, the capacity to buy or lease quota does not always solve the by-catch problem in mixed New Zealand fisheries. To deal with this problem, the government pays the fishermen 50 per cent of the market price for those species for which they do not have quota. This provides an incentive to land the catch once it is caught, while not encouraging the search for such species. The system also allows for up to 10 per cent over quota in a season but the extra landings are deducted from the following season's quota.

Clark et al. (1988) conclude that the New Zealand scheme can be judged to be successful. It effectively addresses both the economic and biological aspects of fishery management but it will continue to be the subject of refinement in the light of experience.

The Icelandic quota management system

In 1984, quotas for cod and other demersal species were allocated to Icelandic fishing vessels according to catch records for the previous three years. Quotas were not divisible nor could they be removed from vessels, except when the vessel was wrecked or sold abroad. Concentration of quota holdings was only possible by buying vessels, and some companies bought old boats for wrecking in order to add the quota to their own vessels. Quota leasing was allowed from 1984 (Eythorsson, 1996) and by January 1991 the system was liberalized and quota shares were allocated permanently without any time limits. Quotas are now divisible; they can be separated from vessels and transferred to other vessel owners, either by transfer of permanent shares or by leasing for one year only. Annual vessel quotas were initially issued free of charge. Under the 1990 legislation, the Ministry collects fees for quotas to cover the cost of monitoring and enforcing the ITQ regulations (Arnason, 1993).

The ITQ system was originally suggested by the vessel owners. The crew members had been more reluctant and their organizations have never accepted the transferability of the quotas. The political parties have not been united on the quota issue either and, indeed, all the relevant laws have been passed by narrow margins (Helgason, 1991).

Transfer of quotas by leasing is a very common practice in Iceland. About one-third of all market transactions are by leasing, an increase of 100 per cent since 1991. There are three main categories of leasing:

- *Quota exchange*. Exchange of quota for different species between vessels may not be leasing in the strict sense but for species such as redfish, plaice and Greenland halibut, there is probably no real leasing market apart from quota exchange. Prices for these exchanges are usually fixed in cod equivalents (which automatically determines the leasing prices for other species).

- *Contract fishing and quota pooling.* These involve contracts between vertically integrated fishing companies with quota holdings, and owners of vessels with little or no quota of their own.
  The boat owners are obliged to deliver the fish to the company
  fish factory and receive a fixed price for the catch which is only
  50–60 per cent of the price obtainable for such fish in auction
  markets. The difference is the quota least price.
- *Direct leasing.* Boat owners lease quota at a market price directly
  from the owner or through a mediator. These boats are free to
  deliver fish where they please. Direct leasing is a solution for by-
  catch problems or to prolong the season for boats short of quota.
  High leasing prices are, however, making this system uneconomic
  for the small-scale fishers. Lease prices for all species have risen
  since 1991 but for the two most important species on the market,
  cod and shrimp, the increases have been dramatic (see page 85).

High quota prices can hardly be explained by increased efficiency,
especially in a situation of stock decline. Arnason (1995) notes that
in the initial phase of ITQ management there may be an increase
in the market value of catch quotas due to the use of excessive
fishing capital that has become practically worthless. The short-
term effect will, according to Arnason, be corrected as soon as the
level of fishing capital reaches a new equilibrium. Eythorsson (1996)
has some doubts about this explanation. He says that the high cost
of cod leases can also be explained by other factors. Cod is the
most abundant and valuable species in coastal waters. Though the
boats try to catch other species, there is always a substantial by-
catch of cod. Fishers buy quota to match this by-catch rather than
dump the cod. Hence, the rapid rise in cod quota prices may be
partly explained by this situation. There is also strong demand for
quotas by local municipalities wishing to create local employment.
As there were no signs of a development towards an equilibrium
between catch capacity and resources, the Icelandic government
instituted a cost buy-back programme in 1994 to remove vessels
from the fleet, but it is too early to judge the effects of this programme on cod leasing prices. Hence, despite the introduction of
ITQs, there is a return to the vessel buy-out method which has not
been very successful in other countries in the past.
As a growing proportion of the Icelandic TAC is being transferred through the high priced leasing market, a large number of
fishermen and crew members are getting lower incomes than before.

The lowering of incomes led to a bitter industrial dispute in 1994 and 1995 between the Icelandic fishermen and the government when the ITQ system and quota profiteering practices were called into question. Eventually, agreement was reached on quota prices. Eythorsson (1996) says that it is too early yet to judge the influence of this agreement on the quota market transactions, but he adds that the strikes are only one of several signs of conflicts related to the ITQ system in Iceland. All political parties now have disagreements about the programme.

The ITQ system has also given rise to problems for local communities. In the early 1980s, 10–20 per cent of the fishing industry was under municipal ownership. Now this figure is reduced to 5 per cent and there is high unemployment in the areas which have lost their quotas.

Eythorsson concludes that privatization of the fishing rights in Iceland is a social experiment involving high stakes. The most significant result to date is a massive redistribution of wealth and income. The winners are the big quota owners who can obtain high annual returns from their new capital. The losers are the smaller fishermen and the fishing crews who have been thrown into a market where eventually only the most efficient companies have a choice to catch the fish.

Individual vessel quotas (IVQ) in the British Columbia halibut fishery

The Pacific halibut fishery, one of the oldest fisheries on the west coast of North America, is exploited commercially by both the US and Canada. Halibut fishery vessels average between 40 and 60 feet in length and are used in other fisheries such as salmon and herring. Most of the firms processing halibut handle a variety of other fish species. Prior to the introduction of the IVQ programme in 1991, the majority of British Columbian-caught halibut was sold frozen. Now almost all is sold fresh.

In the pre-IVQ days, the halibut fishery operated under a limited entry programme with 435 licensed vessels (Casey et al., 1995). However, the fleet capacity increased dramatically during the 1980s through more intensive crewing, introduction of electronic gear, more efficient circle hooks and automatic baiters. By 1990, the fishery season had been reduced to six fishing days per vessel from a 60-day season in 1982. At this time it was evident that the existing management approach was not working. Fishermen fished in hazardous conditions continuously up to 24 hours per day, and as the 'race-to-fish Derbies' became shorter, the incidence of vessel sinkings, injuries and loss of life became more common. Fish quality was reduced by improper handling, processing gluts and the necessity to freeze and hold the catch over most of the year.

Following discussions in 1988/9, during which proposals for alternative management programmes, including IVQs, were largely supported by boat owners but ardently opposed by processors and crew members' unions, the Ministry of Fisheries adopted an IVQ programme on a trial basis for the 1991 and 1992 seasons. The key features of the trial programme were as follows:

- Fishing to remain open over the eight-month season from March to October.
- Each individual licensed vessel to receive an allocation specified as a share of the TAC. Shares were assigned so that 70 per cent were based on the vessel's performance between 1986 and 1989, the remainder being associated with the vessel's length. Disputes were resolved by an appeal board.
- An observer company and a team of fishery officers were hired to monitor and enforce the programme. The system is funded by the fishermen. An observer validates weights of fish caught. Stiff penalties for violations are imposed, including forfeiture of quota.
- During the first two years, quotas were not allowed to be transferred. In 1993, limited transferability was allowed. The restrictions on transfers have limited fleet consolidation. Nevertheless, 74 licensed vessels are no longer fishing for halibut.

The effects of the IVQ programme have been significant:

- Halibut fishing is now spread over eight months compared with six days previously. Over 90 per cent of fish are sold fresh resulting in higher ex-vessel prices.
- There has been a transfer in market share from large traditional firms to smaller firms specializing in halibut, and an increase in the number of firms purchasing halibut.
- There has been a reduction of about a third in total crew numbers. More than half of this reduction has been in vessels still active in the fishery, the remainder being in retired vessels. Those remaining are working longer seasons.

A vote of licence holders in December 1992 showed over 90 per cent in favour of continuing the ITQ programme. The programme is, however, not without its criticisms which focus mainly on the reduction of the number of people employed and the windfall profits accruing to halibut quota holders as the increased profitability of the fishing is translated into higher market values for quotas.

Canadian Bay of Fundy – St Lawrence herring programme

During the early 1970s, the Bay of Fundy herring fish stock collapsed. A number of measures were introduced by the Canadian government involving the setting up of a cooperative to manage an individual quota system. Enforcement proved well nigh impossible and in 1983 a new individual quota programme – The Bay of Fundy–St Lawrence Programme – was substituted for the previous one. This new programme emphasized fleet size reductions whereas the original programme had been designed to assist the existing fleet organize its marketing. Where the first issued temporary non-transferable quotas, the second awarded ten-year transferable quotas.

The enforcement environment for this was as unpromising as for the earlier one. The introduction of an audit system using information derived from landings and processor documents was frustrated as a result of collusion between buyers and sellers. The programme has continued to be plagued by under-reporting. There has also been serious conflict between fishery officers and fishermen at quaysides. Available evidence suggests that the actual landings in the Bay of Fundy region in 1984 were 1.8 times the size of the TAC and that cheating has continued (Muse and Schelle, 1989). Reductions in the size of the fleet have not been as large as anticipated. A fisherman who can expand his operations through relatively riskless under-reporting has little incentive to buy additional quota. Enforcement, therefore, seems to be the kernel of a successful ITQ programme in this area and this can be very expensive.

Experiences with individual quotas generally

An ongoing OECD study indicates that the results of ITQ management programmes have been mixed. For the most part, ITQs have been effective in limiting catch at or below the TAC levels. There have, however, been recurring problems in some fisheries of catches exceeding quotas. Insufficient monitoring, high prices and collusion between buyers and sellers have been the causes of the problems. Canadian Atlantic ground fish stocks, particularly cod, have experienced continued decline under ITQs. These declines began prior to the introduction of ITQs but the system has not been able to reverse them. Excessively high TACs have also been fixed in some cases so that stocks have continued to decline even where catch stayed within the TAC limits.

Several fisheries use additional regulations such as minimum net size, closed seasons and areas, and gear restrictions in conjunction with ITQs. In general there is nothing to show that such added regulations have resulted in superior resource conservation.

The introduction of ITQs was also expected to eliminate the race-for-fish and consequent problems. Such elimination has not been universal. This is probably due to the fact that the fisheries could be closed as soon as national quotas were caught even if individual quotas were not filled. Indeed, most of the fisheries where a race-to-fish has persisted have used time or area closures alongside ITQs.

CONCLUSION

The main conclusion to be drawn from experience to date in applying ITQs is that they do not provide the panacea suggested by their earlier advocates. While, in theory, the conferring of property rights on individual fishers should facilitate resource protection and conservation, in practice, this has not always happened. Commenting on this problem, Hanna et al. (1995) say that:

- Property rights are a necessary but not a sufficient condition for resource sustainability. If assured access to future benefits from resource use is absent, no incentive exists to limit current use.
- The giving of property rights to fishermen does not always ensure reasonable sustainability. Policy must also consider the context in which such rights are placed and the extent to which they are enforceable.

However, despite the many difficulties which have arisen, the OECD study mentioned above indicates that ITQ systems can and have worked in a number of fisheries. There are a number of such programmes in operation in the EU which have been worked out informally between governments and producer organizations. In many cases, these quotas are not transferable, but despite this, swaps and deals of different kinds are occurring.

Because many of the world's fisheries have been exploited to the point of near extinction, traditional systems of management have to be drastically changed. Despite the drawbacks associated with ITQ systems, we think that there should be a concerted movement towards these systems within the EU, starting with pelagic fisheries and moving on later, in the light of experience, to the more difficult ground fisheries.

The following general rules would help alleviate some of the problems which have occurred with ITQs:

- Quota systems should be operated by local producer organizations (LPOs) in cooperation with fisheries departments who would have statutory powers to:
  - make rules regarding transferability;
  - enforce these rules; and
  - regulate prices for quota sales and leases.
- A quota holder should not be allowed to sell or lease quota directly to another fisher. As with EU agricultural milk quotas, fish quotas should only be leased or sold to the LPO at fixed prices and the LPO should in turn sell or lease these quotas to other fishermen or to new entrants.
- Quotas should as far as possible be ring-fenced within designated areas. This, of course, will not be entirely feasible. However, rules should be drawn up to reserve a proportion of the TAC for local fishermen.
- All fish buyers and exporters should be licensed and enforcement should, as in New Zealand, be through completed documents from fishermen and fish buyers. 'Over the side' sales at sea should be policed by naval vessels.
- To prevent high grading, all landed fish should be purchased at market prices, but at the end of each year a levy of 50 per cent of the sale price should be imposed on all over-quota catch. It may be possible to solve this problem to some extent by short-term leasing if flexi-quota is available.
- In mixed fisheries, mixed quotas should be experimented with in an effort to solve the by-catch problem.

NOTES

1. The authors wish to acknowledge the assistance and valuable comments received from the following people:

   Dr Deirdre O'Keefe, Department of Education, Dublin.
   Professor Jim Crutchfield, Natural Resources Consultants, Seattle, Washington.
   Professor William E. Schrank, Memorial University of New Foundland, for the supply of references on Canadian fisheries.
   Professor Torbjorn Trondsen of the Norwegian College of Fishery Science, University of Tronso, for information on Norwegian management systems.
   Professor J. O. S. Kennedy, School of Economics, La Trobe University, Melbourne, and Mr David Campbell, Principal Research Officer, Australian Bureau of Agricultural and Resource Economics, Canberra, for advice on the working of the Australian ITQ system.
   Dr Paul Hillis, Marine Institute Fisheries, Research Centre, Abbotstown, Co. Dublin, for references from all over the world on ITQs.
   Mr Pat Keogh, Mr Alex Heskin and Mr Dominic Rahan of the Irish Sea Fisheries Board who supplied information in ITQs in a number of countries and commented on early drafts of the paper.

   None of the above is responsible for the final version of the paper. Any errors or omissions are the sole responsibility of the authors.
2. Some countries' fleets have outstripped the capacities of their fishing grounds to the extent that Iceland and the EU could cut their boats by 40 per cent and Norway by two-thirds and still catch as many fish as they do today.
3. Hardin's 'tragedy of the commons' takes the example of a pasture in common ownership where each herdsman concludes that the only sensible course for him is to add another animal to his herd, leading inevitably to the destruction of the pasturage. 'Freedom of the commons brings ruin to all.'

BIBLIOGRAPHY

Arnason, R. (1993) *The Icelandic Individual Transferable Quota System*, paper read to Conference on Innovations in Fishery Management, Norwegian School of Economics, Bergen, May.
Arnason, R. (1995) *The Icelandic Fisheries, Evolution and Management of a Fishing Industry*, quoted in Eythorsson (1996).
Casey, K. E., Dewnes, C. M., Turris, B. R. and Wilen, J. E. (1995) 'The Effects of Individual Vessel Quotas in the British Columbia Halibut Fishery', *Resource Economics*, 10: 211–30.
Clark, I. N., Major, P. J. and Mollett, N. (1988) 'The Development and Implementation of New Zealand's ITQ Management System', *Marine Resource Economics*, 5(4): 325–49.

The Economist (1994) 'The Tragedy of the Oceans', 19 March.

Eythorsson, E. (1996) 'Theory and Practice of ITQs in Iceland', *Marine Policy*, May.

Geen, G. and Nayar, M. (1988) 'Individual Transferable Quotas in the Southern Bluefin Tuna Fishery: An Economic Appraisal', *Marine Resource Economics*, 5: 365–87.

Grafton, R. Q. (1996) 'Individual Transferable Quotas and Canada's Atlantic Fisheries', in Gordon D. V. and Munro R. (eds), *Fisheries and Uncertainty – A Precautionary Approach to Resource Management* (Calgary: University of Calgary Press).

Hanna, S., Folkle, C. and Maher, K. G. (1995) 'Property Rights and Environmental Resources', in Hanna, S. and Munasinghe M. (eds), *Property Rights and the Environment* (Washington, DC: The Beijer International Institute of Ecological Economics, Sweden, and the World Bank).

Hardin, G. (1968) 'The Tragedy of the Commons', *Science*, 1612: 1243–8.

Helgason, T. (1991) *The Icelandic Quota Management System, A Description and Evaluation* (Science Institute, University of Iceland, March).

Huppert, D. D. (1989) 'Comments on Clark, I., Major, P. J. and Mollett, N., "The Development and Implementation of New Zealand's ITQ management system"', in Neher et al. (1989), pp. 146–9.

Kaitala, V. T. and Munro, G. (1995) 'The Management of Transboundary Resources and Property Rights Systems: The Case of Fisheries', in Hanna, S. and Munasinghe. M. (eds), *Property Rights and the Environment, Social and Ecological Issues* (Beijer International Institute of Ecological Economics, Sweden, and Washington, DC: World Bank).

Libecap, G. D. (1989) 'Comments on Anthony D. Scott, "Conceptual Origins of Rights Based Fishing"', in Neher et al. (1989), pp. 39–45.

MacGillivray, P. B. (1990) *Assessment of New Zealand's Individual Transferable Quota Fishery Management: Economic and Commercial Analysis* (Report No. 75, Canadian Department of Fisheries and Oceans).

McLysaght, C. (1991) 'Legal Aspects of the CFP Review', in *Review of Common Fisheries Policy* (Report of the Advisory Group set up by the Minister for the Marine, Department of the Marine, Dublin), pp. 117–34.

Muse, B. and Schelle, K. (1989) *Individual Fishermen's Quotas: A Preliminary Review of Some Recent Programs* (CFEC 89–1, Alaska Commercial Fishery Entry Commission, February).

Neher, P. A., Arnason, R. and Mollett, N. (eds) (1989) *Rights Based Fishing* (Dordrecht: Kluwer Academic).

Panayotou, T. (1989) 'Comments on K. Ruddle, "The Organization of Traditional Inshore Fishery Management Systems in the Pacific"', in Neher et al. (1989), pp. 86–93.

Rettig, R. B. (1989) 'Is Fishery Management at a Turning Point? Reflections on the Evolution of Rights Based Fishing', in Neher et al. (1989), pp. 47–64.

Ruddle K, (1989) 'The Organisation of Traditional Inshore Fishery Management Systems in the Pacific', in Neher et al. (1989), pp. 73–83.

Scott, A, (1989) 'Conceptual Origins of Rights Based Fishing', in Neher et al. (1989), pp. 11–38.

Thomson, D. B. (1996) 'Fish Production and Markets – Global Perspective', paper presented to Conference 'Towards a Market Policy for Ireland', Connemara Coast Hotel, Co. Galway, Marine Institute, February.

Townsend, R. E. and Pooley, S. G. (1995) 'Distributed Governance in Fisheries', in Hanna, S. and Munasinghe, M. (eds), *Property Rights and The Environment: Social and Ecological Issues* (Washington, DC: The Beijer International Institute of Ecological Economics, Sweden, and the World Bank).

Trondsen, T. and Angell, J. (1992) *Regional Enterprise Share Quota Management System, the Case of Norway*, paper prepared for the Annual Conference of the European Association of Fishery Economists, Salerno, Italy, April.

7 Regulation and Representation: Institutional Challenges in Fisheries Management[1]

Knut H. Mikalsen

INTRODUCTION

Legitimacy and compliance have gradually become key concepts in fisheries management, sometimes surpassing efficiency and conservation as conspicuous catchwords in debates on management policies (Jentoft, 1993; Felt, 1990). There is a growing recognition – among managers, scientists and politicians – that no management scheme will work unless it enjoys the support of those whose behaviour it is intended to affect. Legitimacy, in this sense, has to do with compliance, with decisions and policies that conform to, or approximate, the values, standards and expectations of those affected (Beetham, 1991: 11). One of the most pressing problems in fisheries management, then, is how to ensure 'grass-roots' approval of decisions that may actually harm large segments of the industry.

The standard solution to this has been *representation*, the participation of interest (or user) groups in the making of management policies. Corporatism (Schmitter, 1974), policy communities (Pross, 1992), iron triangles and policy networks (Jordan, 1981) are terms that have been used more or less interchangeably to conceptualize structures of state–group relations within distinct policy areas. These structures tend to be highly selective in that only user-groups proper are included. Groups and interests that may be affected by management policies in more indirect ways are typically excluded – without this being considered illegitimate or even undemocratic. Democracy, in this context, is tantamount to institutional arrangements that allow those *directly* affected by public decisions to have a say.

Within fisheries management, recent developments have led to

100

the questioning of state–group relations and established decision-making structures. Should, for example, user-group participation be a central value at all? A Norwegian fisheries scientist may not have been speaking only for himself when he stated '. . . I'm certainly for democracy, but when it comes to fisheries management I'll support dictatorship'. There has, however, been little support for more centralization. Rather, there have recently been repeated calls for a more open and less exclusive process. In Norway, ethnic minorities (e.g. the Saami) and environmental groups have demanded a seat on advisory committees, and arguments in favour of consumer representation in management policy-making have been heard. In this sense there are calls for more democracy rather than less.

Corporatism, the argument goes, is tantamount to letting the fox into the henhouse. The opinions of users (i.e. fishermen) need to be balanced against the interests of the broader community. Besides, corporatist arrangements weaken democratic accountability as important decisions are increasingly made outside political institutions proper. Assertions such as these should be familiar to most students of public policy-making, and they indicate that there may, indeed, be problems and dilemmas associated with user-group participation in fisheries management. What are these and how can they be dealt with? Is there a case for extending representation beyond user-groups proper, and how could this be achieved without undermining the legitimate claims of users to be influential actors?

THE PRESENT REGIME: CENTRALIZED CONSULTATION

With regard to organizational structure, fisheries management in Norway is characterized by a combination of centralized control and extensive consultation. The former is secured through legislation vesting the ultimate authority to manage in the hands of central government agencies; the latter is achieved through corporatist arrangements facilitating the participation of organized interest groups. The national level has by and large been regarded as the natural locus of user-group participation in management. It follows from this that user-group involvement does not necessarily entail decentralization. Apart from the Lofoten fishery (Jentoft and Kristoffersen, 1989), there are few, if any, cases of formalized and 'official' systems of local management. Rather, the present system is largely a trade-off between hierarchical control and corporatist participation, rule-making and log-rolling, science and politics.

At the core of this structure is the so-called *Regulatory Council*, an advisory body to the Minister of Fisheries including representatives from most segments of the industry as well as from government and science. As such it is part of an intricate network of committees, and of a long-standing tradition of functional representation in Norwegian politics (Rokkan, 1966; Nordby, 1994). Its main role is advising the Minister on the allocation of the TAC (total allowable catch) of different species among different groups of fishermen or segments of the fleet such as offshore and inshore. By its composition, the council is a mixture of the purely professional and overtly political – a body of knowledgeable people with interests at stake.[2] The system, within which the Council is a major player, has been labelled 'centralized consultation' (Hoel et al., 1996) and its basically top-down structure is not entirely accidental.

First, it is both common and reasonable to regard the fish stocks within the 200-mile zone as *common, or national, property*, as part of our national heritage. Their management can thus be said to be the responsibility of national, or nationwide, institutions, i.e. government agencies. The fish stocks, the argument goes, should be managed on behalf of the national community, and they are therefore best left to institutions which, in principle at least, act with a view to the public interest. Besides, the fact that most fisheries extend well beyond local and regional boundaries in itself justifies a centralized regime. In this context, legitimacy and support is a question of conferring with the industry rather than with regional agencies or local-level associations.

Second, regulatory measures *do confine, sometimes severely, the economic independence of fishermen*. And they raise the problem of allocating scarce goods (e.g. licences and quotas) among users claiming, often rightly so, the same rights and needs. In this context, management decisions and policies will rarely be accepted without prior consultation. Further, scarcity implies 'cruel' choices whose legitimacy will be strengthened if civil service norms of rule-following, impartiality and equal treatment are observed. And it *could* be argued that consultation requires selective representation, and that the norms and virtues referred to above can best be observed through standardized rules and centralized decision-making.

Third, management has *intensified traditional conflicts* within the industry and created new areas of friction and dispute. These are not easily handled at the subnational level as regional and local institutions will lack the political authority to intervene effectively. Government intervention is often necessary in order to solve the conflicts over distributional issues entailed by regulatory schemes; consultations with affected interests are a necessary precondition to make decisions legitimate and binding.

Fourth, more than 80 per cent of Norwegian fisheries utilize stocks shared with other countries – notably Russia and the EU. Fisheries management then, is to a considerable extent a question of *international negotiations* – and these are, constitutionally speaking, the exclusive domain of central government.

DILEMMAS OF USER-GROUP PARTICIPATION

There are, in other words, 'rational' explanations for what has been labelled 'centralized consultation' in Norwegian fisheries management. Besides, the system is rooted in history and tradition, its legitimacy sustained by the fact that it does produce viable compromises. Any advantages, however, will have to be weighed against some of the problems and dilemmas entailed by the close relationship between government and industry in a management context. The first and most familiar pertains to the *relationship between interests and influence*. It is a central assumption in democratic theory that groups directly affected by government decisions should have a say in the making of them, in our case decisions on how quotas are set and licences allocated. Participation is a fundamental criterion of procedural democracy (Dahl, 1986), and corporatist structures provide an institutional arena for expressing preferences and demands. Corporatism may, on the other hand, facilitate user-group influence at the expense of government control – putting special interests before the public good (Anderson, 1979). Giving industry the power to (co-)manage is sometimes tantamount to, as the Norwegian saying goes, letting the goat guard the oatmeal sack, as there is a strong temptation to capture immediate benefits (i.e. larger quotas), while disregarding the long-term impact on the stocks.

One could, of course, turn the argument on its head by saying that the goat, in particular, has a genuine interest in stretching the contents of the sack as far as possible – and that it will act accordingly. Whatever the case, the general conception has been one of an industry easily tempted and of a government only too willing to oblige – with dire consequences for the resource. The obvious objection to such a conclusion is that the TAC is decided by scientists – without interference from government or interest groups. Scientific assessments, however, are tenuous at best, and the lack of reliable data makes it tempting to solve problems of allocation by increasing the overall quota. Besides, resource conservation is only one of several objectives of Norwegian fisheries management. As such it has to be balanced against economic efficiency, regional development and professional protection (Mikalsen, 1985).

In an industry populated by diverse interests and competing groups, the present system of centralized consultation raises *the issue of representativity*, the question of whose interests are being served or accommodated by present arrangements. Historically, the Fishermen's Association has enjoyed privileged access to government – in management as well as in other policy areas – while processors and plant workers have lacked political 'clout' (Mikalsen and Sagdahl, 1982; Hallenstvedt, 1982). One reason for this is that management, in particular, has been defined as an issue that mainly pertains to harvesting. This is clearly reflected in the composition of the Regulatory Council where the association has more representatives than any other group or institution. However, the fact that the industry in general, and fishermen in particular, are represented in policy-making does not necessarily mean that all interests are given a fair and proper hearing. The aggregation and coordination of demands within the Fishermen's Association is a case in point. Due to the heterogeneous nature of this organization, aggregation has always been difficult and time consuming. It is basically a fragile coalition of offshore and inshore groups, of boat owners, crew members and independent operators – groups whose administrative capacity and political clout differ considerably (Jentoft and Mikalsen, 1987). Consensus is hard to reach, and opposition to management decisions and policies is frequent, particularly among inshore fishermen claiming their interests are being ill served by present arrangements.

The point is that representation on councils and committees, in a conflict ridden industry like the fisheries, does not guarantee that demands and decisions are 'representative'. This, of course, is not just a problem for the industry, but also for government in its role as mediator. For the organization(s) involved, the close relationship with government may affect their role *vis-à-vis* ordinary members. To the organizational leadership, participation means obligations and responsibilities, and the risk of being coopted by the state. As a consequence, the organizations involved may come to play the role of a government agency rather than that of a genuine and effective interest group.

A third aspect, or dilemma, of centralized consultation pertains to the classical problem of *balancing special interests against public concerns*. Corporatist arrangements imply selective representation in that some groups are considered to be more affected than others by a given policy, and hence entitled to a seat at the table. As such they are highly conducive to what Theodore Lowi (1969) has labelled interest group liberalism whereby, among other things, formal procedure is weakened by informal bargaining. Fisheries management, for example, has long been based on the presumption that the making of regulatory policy is the exclusive domain of the industry, even of a particular segment of it (harvesting), in close cooperation with government. The fact that decisions may affect wider groups has largely been neglected – unless it has been considered politically advantageous. For example, union officials always take great care in emphasizing the regional significance (and hence, the public interest nature) of the fisheries when government subsidies are justified. This aspect is, however, seldom acknowledged when 'affected interests' are defined and participants selected. In this context, a commendable policy is, almost by definition, one to which those directly involved can agree, or one that is favoured by a winning coalition. There is, in other words, little scope for executive discretion or ministerial fiat. In this sense, consensus and legitimacy rather than efficiency and governance are core values of management procedures.

The problem, however, is that other groups can claim a stake in management decisions, and that they increasingly do so. *Consumers*, for example, have a legitimate interest in the availability and quality of fish products, which may be affected by management practices. The average citizen may not know, nor much care, about the difference between TACs and IQs, but the moment fresh fish is no longer available he or she is bound to start asking why. *Regional and local authorities*, particularly in northern Norway, have come to take a keen interest in the effects of management policies on employment opportunities and the welfare of local communities. The resource crisis of 1989–90, in particular, made elected officials acutely aware of the importance of a viable fishery to local economies, and there has since been much talk about the prospects of sharing responsibility for management. *Environmental groups*, for their part, are speaking for the fish in targeting the preservation of stocks and the balance of highly vulnerable ecosystems. Finally, *ethnic groups* (e.g. the Saami) argue that management practices should acknowledge historical rights and the need to protect traditional ways of life, and that this, among other things, calls for a more decentralized decision-making structure.

The two latter groups have recently been recognized as legitimate participants, one as a full member of the Regulatory Council, the other as observer. They can, as yet, hardly be considered core members of the 'management community'.[3] The point, however, is that issues of fisheries management are increasingly catching the public eye, and that this seems to be a world-wide trend. In North America, there is less talk of user-groups and more of 'stakeholders', a term which broadens the scope of affected interests considerably. In international organizations, e.g. the UN, fish stocks are being referred to as 'the common heritage of mankind', and there is a quest for management regimes that will facilitate a better trade-off between the legitimate interests of the fishing industry and the long-term concerns of the public at large.

TOWARDS BROADER PARTICIPATION?

There are, in other words, pressures for broadening representation and participation in management decision-making, perhaps even for moving decisions and debates out of the 'back rooms' and into more public and transparent arenas. Public attention may, of course, prove temporary, generated by the present crisis and political turmoil in domestic and international fisheries. It may, on the other hand, also reflect genuine concerns, and a growing perception of fisheries management as too important to be left to industry and government alone. This is a point argued by environmental groups in particular, and their partial inclusion in the policy community (in Norway) is a recognition of their concerns as both legitimate and important.

There are, however, two types of decisions involved here. Environmental groups have, so far, paid more attention to the setting of overall catch levels (i.e. TACs) than to the allocation of quotas among user-groups. The latter issue has largely been defined as almost an 'internal' one. This applies, in particular, to the allocation between offshore and inshore which is settled through the so-called 'quota ladder'.[4] For the government this is, of course, politically convenient as it will not be held responsible for one of the most controversial allocations in Norwegian fisheries management. For the Fishermen's Association, however, the allocation issue has led to internal conflicts and threatened to rip the organization apart. Besides, it could be argued that the allocation of quotas between offshore and inshore affects others than those directly involved. Allocation affects how quotas are caught, which is of some consequence for the future state of the stocks, and this is something we all have a stake in.

The question, then, that needs to be put is whether allocation in this sense should continue to be regarded as an 'internal' issue or whether the Minister should make an independent decision based on broader consultations. Changes along these lines could free the Fishermen's Association of a burdensome task, and may therefore seem attractive to its leadership. It would, however, move an important issue beyond industrial control, and reduce the political significance of the Association. In this sense, internal strife is perhaps the price one has to pay for political status and influence.

While producer groups, through the Fishermen's Association, have had a considerable influence on allocation, their views do not count as much when stock assessments are made and TACs set. This is more or less the exclusive domain of science. Should it continue to be so? There is, among scientists, a growing appreciation of how the complexity of the biological environment creates an enormous information problem that seems to undermine any pretence of scientific certainty. Add to this the observation that scientists work within a social environment and that their judgements may not be immune to political pressure (Finlayson, 1994), and you have the origin of recent debates on the role of numerical analysis in fisheries management (Wilson and Kleban, 1992) as well as on how user-group knowledge can be utilized for management purposes (McCay and Acheson, 1987; Dyer and McGoodwin, 1994). The question, then, is whether user-groups should be more involved in generating the knowledge on which TACs, for example, are based. There is a challenge here in designing institutional arenas for a more systematic and elaborate exchange of observations and views between fishermen and scientists with the aim of improving the quality of stock assessments.

Recent changes in the management structure of Atlantic Canadian fisheries may provide an ample illustration. A few years ago, a new institution was created to deal with issues of stock assessments and the setting of TACs – 'The Fisheries Resource Conservation Council'. Even though representation is restricted, the Council operates as a forum where managers, scientists, fishermen and processors can discuss matters of common concern. One objective is to involve industrial representatives in generating information about the state and development of stocks (Government of Canada, 1994). The Council works, in part, through public hearings where representatives of interest groups are given the opportunity to comment on initiatives from government and science – and to convey their views on stock assessments and management proposals. During 1993, for instance, the Council conducted more than twenty public hearings along the coast of Atlantic Canada. Together with updated reports from scientists on the state of the stocks, these hearings formed the basis for the Council's recommendations on quotas and management practices to the Minister of Fisheries and Oceans.

FROM CORPORATIST STRUCTURES TO INDEPENDENT BOARDS?

Fisheries management is, as already noted, not just a question of setting TACs on the basis of scientific recommendations and user-group advice. It is as much about allocating quotas and licences, enforcing sanctions and designing principles and institutions for what are basically political decisions. Considering the fact that the efficiency of management measures hinges on compliance, and compliance on legitimacy, there may indeed be a strong case for user-group participation in management decision-making. Representation not only provides user-groups with an opportunity to shape policy, it also makes them responsible for policy and increases the likelihood of cooperation and support in implementation.

Assuming that public policies are often contested, not only for their actual content, but also on purely procedural grounds, legitimacy and compliance may not hinge entirely on direct representation. Where representation is not feasible, as when users are not organized, or perhaps not even desirable, as when the number of legitimate participants is too large to be accommodated within representative structures, keeping affected groups at arm's length may be the only way to secure broad support for programmes and decisions. If a decision or policy is considered to be the result of a particular procedure that entails, say, impartial and knowledgeable judgment, it may be considered legitimate – and hence supported and complied with – even if it does not directly accommodate the values and expectations of those affected. What we are talking about here, in other words, is achieving legitimacy through *impartiality and independent judgment*. Both 'methods', *representation* and *judgment*, pertain to procedure rather than content. They are different, however, in that the former entails user-group involvement and close, formalized cooperation between government and the governed, while the latter implies delegation of management tasks to agencies or committees of knowledgeable and (presumably) independent people.

What would such a system, modelled on the North American regulatory agency or the British quango, look like in fisheries management? Let me illustrate this by a brief description of an institutional reform recently proposed, but not yet adopted, in Atlantic Canadian fisheries. The main thrust of these reforms is the proposal of replacing a complex structure of advisory groups (almost forty in Atlantic Canada alone) with a 'Fisheries Board' of individuals 'knowledgeable about and with experience related to the industry', but with no direct financial interest in it. This board would also take over responsibilities and powers currently exercised by the Minister of Fisheries and Oceans such as the allocation of licences and quotas. They would also, according to the proposal, replace the criminal courts in enforcing sanctions for violations of management rules (Government of Canada, 1993; Parsons, 1993).

What has been proposed, in other words, is the establishment of an independent body with the power to make binding decisions on licensing, quota allocations and enforcement, decisions that are to be made on the basis of public hearings with affected interests and a policy framework set by the Department of Fisheries and Oceans (DFO). What we are dealing with here is a model where real decision-making power is delegated to an independent and presumably neutral body, with no direct and formal representation of user-groups. This is, in essence, a solution that seeks to combine central governance, 'functional' authority and democratic control: goals and policies are to be set by DFO, while allocations of licences and quotas are to be decided by an independent body of 'experts' on the basis of public hearings that will include a broader selection of groups and interests than does the current system. If adopted, it would bring the management of fisheries more in line with regulatory practices in other Canadian industries where independent, quasi-judicial bodies seem to be the norm. These are required to conduct their business in a judicial-like manner, framing their decisions to fit policy guidelines rather than precise legislative standards.

WHY INDEPENDENT AGENCIES?

The creation of independent agencies at arm's length from both government and user-groups, could be justified on several grounds. To begin with, it may be tantamount to removing an issue, or issue area, beyond the realm of politicking, i.e. from processes that rely 'too much on partisanship, compromise, and expediency, and not enough on fairness and hard economic facts' (Kernaghan and Siegel, 1991: 229). In this sense, the creation of management boards would put the oatmeal sack beyond the reach of the goat, providing a better balance between user-group demands, government priorities and scientific recommendations in management decision-making. That said, there are certainly limits to the depoliticization of the allocative issues involved. In almost any industry, and certainly within Norwegian fisheries, a board structure would operate in an environment where resources are scarce and where, traditionally, outcomes have been decided by political clout rather than formal rules and technical information. Members of a board will have to confront cases where the livelihoods of individuals and groups are at stake, and where decisions are unlikely to be accepted on the ground that they have been reached through informed and impartial judgment.

Second, independent agencies may provide more openness within this particular area of public policy by broadening 'representation' beyond the special interest groups involved. By providing opportunities for those involved to state their case, and others to comment and oppose, public hearings will mean that greater attention is paid to the public interest nature of management decisions. An agency structure would, in short, constitute arenas for the 'sounding out' of interested groups and individuals; it would not be an arrangement for the efficient, and exclusive, representation of selected interests. In principle, this would provide for a more inclusive and open process – contrary to that provided by corporatist arrangements and consultative committees. To the extent that participation is open and inclusive, i.e. a matter of choice rather than one of selection, management boards will contribute to greater fairness and more democracy in management policy-making. Greater openness, however, may contribute to the 'socialization' of management conflicts as more groups are drawn into the fray (Schattschneider, 1960) – making the process considerably more complex and time consuming.

Third, a board structure would facilitate a more 'technical', 'professional' and judicial-like approach to management decision-making – playing down the overtly political and adversarial nature of, say, quota and licence allocation. A board may be composed of people with experience from and expertise in particular segments of the fishing industry, and thus facilitate a more efficient utilization of specialized knowledge in fisheries management. Although there is much to be said in favour of strengthening the role of knowledge and expertise in fisheries management, there is, of course, always the question of who the experts are. What kind of expertise is particularly relevant to, say, the allocation of quotas and licences? Are there other sources of knowledge of the fishing industry than the bureaucracy and the industry itself? To what extent should a keen interest in the future of the industry count as a form of expertise? Questions like these must be addressed, and answers to them will certainly affect the composition of management boards, as will the ways in which members are to be appointed – how and by whom?

There is, however, the classical dilemma between political accountability and regulatory autonomy. On the one hand, the control over appointments and policies will normally reside with the Minister since he or she will be politically responsible for decisions taken by a board. On the other hand, adopting the arm's-length principle implied by the agency model requires that the board be fully independent in deciding individual cases. Ministerial control, which follows from the principle of political accountability, may cause a board to be perceived as just another branch of central government, deflate the morale of its members and undermine their independence. Little ministerial interest and supervision, however, could make a board an easy target for powerful interest groups.

CONCLUSION

Due, in part, to perennial problems of compliance and control, legitimacy has become a key issue in fisheries management. Involving user-groups in policy-making has long been the standard strategy to strengthen compliance and sustain the legitimacy of regulatory decisions. Hence the proliferation, in most public policy areas, of committees and councils – geared to the provision of knowledge and advice as well as to the 'production' of support and compliance. Corporatist arrangements of this kind have made it possible to strike a balance between the 'functional' needs of management bureaucracies (for information and advice) and the democratic aspirations of politicians and interest groups (for support and representation respectively).

Whatever the merits of corporatism, the problem of democracy remains. Why are certain groups granted exclusive representation in policy-making when the decisions taken may eventually affect a wider public? Participation is, of course, often limited for reasons of efficiency: not everyone affected can possibly participate if quick decisions are essential. The problem, however, is that restricting participation is not necessarily about making intelligent and democratic decisions, but rather a strategy for the streamlining of organizational choice. That said, exclusive participation may be easier to justify when a particular decision affects A more than B, than in a situation where a multitude of groups have interests at stake. The problem, however, lies in agreeing on which issues are 'special' and which are 'general' – making the interpretation and definition of issues an essential aspect of politics (Kingdon, 1984).

For reasons already alluded to, this aspect of politics is becoming increasingly apparent in fisheries management. Groups and voices with no ties to the industry as such are challenging prevailing definitions of management issues as well as the exclusive participation of user-groups – raising the more fundamental question of who should have the right to give advice and make decisions. To borrow a phrase from Harrop (1992: 270), there is a certain 'crowding of the policy environment' as new groups are challenging customary practices and the autonomy of the established policy community.

The crucial question, given that there are (practical) limits to the direct representation of interests, pertains to the issue of alternative institutional arrangements in a situation where the concept of 'stakeholder' rather than that of 'user-group' is increasingly being used in defining the proper scope of the policy community. Keeping industry as well as other groups at arm's length by transferring decision-making power to independent boards, as proposed in the Canadian context, could be an alternative here. Opting for institutional change along these lines need not imply that industrial groups and interests are neglected. What it means is that they lose their privileged position as their (economic) interests will have to be balanced against the concerns of other stakeholders through open hearings where the final decision lies with a group of knowledgeable and independent people. As such, a board structure may reduce the scope for political compromise, expediency and lobbying in fisheries management, provide for more transparent decision-making and facilitate the utilization of specialized knowledge.

There is – and this is perhaps the main point of the chapter – a balance to be struck between the legitimate and vested interests of user-groups, whose livelihoods depend on management decisions, and the genuine concerns of, say, environmental groups and other stakeholders – be they ethnic minorities or future generations. Striking this balance is largely a question of creating appropriate institutions. Corporatist structures, prevalent in most management regimes, are based on (immediate) economic interest as the only valid criterion for representation. They may thus be less adequate – and legitimate – given the public interest nature of fisheries management. Independent boards, as outlined in this chapter, are more transparent and inclusive – and have a *public* dimension that is clearly lacking in corporatism. If the legitimacy of management decisions increasingly depends on whether they can be argued and justified in public, there may be much to be said for regulatory institutions that facilitate open debates and public scrutiny without necessarily undermining the need for efficiency in management decision-making.

NOTES

1. Thanks, for comments and discussions, are due to Richard Apostle, Hans-Kristian Hernes and Bonnie McCay.
2. Represented on the Council are the Norwegian Fishermen's Association (5), the National Association of Fish Processors (2), the Norwegian Seamen's Association (1), the Norwegian Union of Plant Workers (1), the Directorate for the Management of Natural Resources (1), the Directorate of Fisheries (2), the Marine Science Institute (1) and the Saami Parliament (1). Two things are worth noticing here: first, the relatively strong position of the Fishermen's Association (five members out of 14), and the fact that the industry as such commands a clear majority (nine out of 14); second, the recent inclusion of institutions

and organizations representing environmental concerns (the Directorate for the Management of Natural Resources) and ethnic rights (the
Saami Parliament).
3. The decision to grant observer status to the National Association of
   Environmental Groups met with vociferous opposition from individual
   members as well as local associations of the Norwegian Fishermen's
   Association. 'We've had enough. The Association should now recon-
   sider its participation in the council,' said a spokesman. Others voiced
   concerns about ' . . . the broadening of participation from organizations
   and groups based on (irrational) sentiments'. There were, at the time
   when the decision was made, fears that public interest groups with
   little understanding of fishing would come to dominate management
   policy-making. Several local and regional branches of the Fishermen's
   Association tried to put pressure on government by passing resolutions
   to the effect that this had to be stopped. See, for instance, *Fiskaren*, 30
   November 1990 and *Sunnmørsposten*, 5 December 1990. Today, the observers seem to have been accepted, and there is general consensus
   among user-group representatives that the new 'members' have acted
   'responsibly'.
4. This is a scheme, or set of guidelines, agreed to by all groups within
   the Fishermen's Association, according to which the relative share of the
   offshore and inshore sectors will vary by the size of the Norwegian TAC.

BIBLIOGRAPHY

Anderson, C. W. (1979) 'Political Design and the Representation of In-
terests', in Schmitter, P. C. and Lembruch, G. (eds), *Trends Toward
Corporatist Intermediation* (London: Sage).
Beetham, D. (1991) *The Legitimation of Power* (NJ: Humanities Press Inter-
national).
Dahl, R. (1986) *Democracy, Liberty and Equality* (Oslo: Norwegian Uni-
versity Press).
Dyer, C. L. and McGoodwin J. R. (eds) (1994) *Folk Management in the
World's Fisheries* (Colorado: University Press of Colorado).
Felt, L. (1990) 'Barriers to User Participation in the Management of the
Canadian Atlantic Salmon Fishery', *Marine Policy*, 14: 345–60.
Finlayson, A. C. (1994) *Fishing for Truth: A Sociological Analysis of Northern
Cod Stock Assessment from 1977 to 1990* (St John's, Newfoundland: In-
stitute of Social and Economic Research, Memorial University of New-
foundland).
Government of Canada (1993) *Fisheries Management: A Proposal for Re-
forming Licensing, Allocation and Sanctions Systems* (Ottawa: Depart-
ment of Fisheries and Oceans).
Government of Canada (1994) *Fisheries Resource Conservation Council:
Building Partnerships in Resource Conservation* (Ottawa: Department of
Fisheries and Oceans).
Hallenstvedt, A. (1982) *Med lov og organisasjon* (Oslo Bergen Tromsø:
Universitetsforlaget).

Harrop, M. (ed.) (1992) *Power and Policy in Liberal Democracies* (Cambridge: Cambridge University Press).

Hoel, A. H., Jentoft, and Mikalsen, K. H. (1996) 'Problems of User-Group Participation in Norwegian Fisheries Management', in Meyers, R. M. et al. (eds), *Fisheries Utilization and Policy: Proceedings of the World Fisheries Congress* (New Delhi: Oxford and IBH Publishing).

Jentoft, S. (1993) *Dangling Lines: The Fisheries Crisis and the Future of Local Communities – The Norwegian Experience* (St John's, Newfoundland: ISER Books).

Jentoft, S. and Kristoffersen, T. (1989) 'Fishermen's Co-management: the Case of the Lofoten Fishery', *Human Organization*, 48: 355–65.

Jentoft, S. and Mikalsen, K. H. (1987) 'Government Subsidies in Norwegian Fisheries: Regional Development or Political Favouritism?', *Marine Policy*, 11: 217–28.

Jordan, G. (1981) 'Iron Triangles, Woolly Corporatism or Elastic Nets: Images of the Policy Process', *Journal of Public Policy*, 1: 95–123.

Kernaghan, K. and Siegel, D. (1991) *Public Administration in Canada* (Scarborough: Nelson Canada).

Kingdon, J. W. (1984) *Agendas, Alternatives and Public Policies* (Boston: Little, Brown).

Lowi, T. (1969) *The End of Liberalism* (New York: W. W. Norton).

McCay, B. and Acheson, J. M. (1987) *The Question of the Commons: The Culture and Ecology of Communal Resources* (Tucson: University of Arizona Press).

Mikalsen, K. H. (1985) *Limits to Limited Entry? License Limitation as a Fisheries Management Tool*, paper presented to the 20th Annual Meeting of the Atlantic Association of Sociologists and Anthropologists, University of Prince Edward Island, Charlottetown, Canada.

Mikalsen, K. H. and Sagdahl, B. (1982) *Fiskeripolitikk og Forvaltningsorganisasjon* (Oslo Bergen Tromsø: Universitetsforlaget).

Nordby, T. (1994) *Korporatisme på Norsk, 1920–1990* (Oslo: Universitetsforlaget).

Parsons, L. S. (1993) *Management of Marine Fisheries in Canada* (Ottawa: National Research Council of Canada).

Pross, A. P. (1992) *Group Politics and Public Policy* (Toronto: Oxford University Press).

Rokkan, S. (1966) 'Numerical Democracy and Corporate Pluralism', in Dahl, R. A. (ed.), *Political Oppositions in Western Democracies* (New Haven, Conn.: Yale University Press).

Schattschneider, E. E. (1960) *The Semisovereign People* (Hinsdale, IL: The Dryden Press).

Schmitter, P. (1974) 'Still the Century of Corporatism?', in Pike, F. and Stritch T. (eds), *The New Corporatism* (Notre Dame: University of Notre Dame Press).

Wilson, J. A. and Kleban P. (1992) 'Practical Implications of Chaos in Fisheries: Ecologically Adapted Management', *Maritime Anthropological Studies*, 5: 67–75.

8 Recent Industry Participation in UK Fisheries Policy: Decentralization or Political Expediency?[1]

Mark Gray

INTRODUCTION

For the purposes of this chapter, fishermen's organizations (FOs) are defined as those exclusively representing the interests of the catching sector. These interests range from informal port and regional associations often preoccupied with local issues (e.g. producer organizations) to sectoral associations concerned with a particular fishing sector (e.g. the Scottish Pelagic Fishermen's Association (SPFA)), and national federations (e.g. the National Federation of Fishermen's Organizations (NFFO)) negotiating management issues with the UK government and Brussels on behalf of the industry.

The UK's centralized fisheries management structure has tended to deny fishermen direct participation in the policy-making process. Following the establishment of the Common Fisheries Policy (CFP) and the consequent elevation of fisheries control to the European level, the distance between FOs and decision-makers appears to be wider. However, mounting pressure from FOs, especially over the last two decades as fishing opportunities[2] and profits[3] have fallen, has been exerted on the authorities to involve them.

There are several reasons for this increased demand for participation. First, fisheries issues have risen up the political agenda in recent years as Members of Parliament with fishing constituencies have frequently voiced their concerns in Parliament over the growing hardships threatening their constituency fishing communities. Second, Eurosceptic MPs have used the perceived failure of the

CFP to support their cause for severing links with Europe and as a result the industry has received greater media attention than in the previous twenty years. The main controversies include the proper role that fishermen should have in management, both as a matter of right to improve the quality of the policy decisions and to add credibility to the management regime.

The purpose of this chapter is to consider the role of the industry in developing UK fisheries policy. Also it assesses whether the government's recent drive to improve the relations with the industry through greater consultation is a recognition of the legitimacy of fishermen's claims to greater participation, or is merely an act of political expediency in the run-up to the 1997 general election.

THE FEDERATIONS' POSITIONS IN DECISION-MAKING UNTIL 1996

In England and Wales, FOs are characteristically small, confined to ports and focusing more on local issues than developing links with other FOs. The lack of coordination and diverse range of interests can make the task of achieving unanimity among its members an arduous one for the National Federation of Fishermen's Organizations (NFFO), set up in 1977 and now representing 50 FOs. However, producer organizations are better structured, and, since 1995, the NFFO has been based on six producer organizations and five regional committees.

Regional and sectoral FOs in Scotland are well organized and highly influential. Just seven FOs form the Scottish Fishermen's Federation (SFF), the Scottish equivalent of the NFFO. These include the Scottish White Fish Producers Association (SWFPA) and the Scottish Pelagic Fishermen's Association Ltd (SPFA), the two largest sectoral organizations in the UK. The SFF was set up in 1973 specifically to negotiate with Brussels on behalf of the Scottish fishing industry.

In Northern Ireland, no single organization represents the entire fishing industry, due largely to the historic rivalry that exists between the ports of Portavogie and Killkeel. Interests are represented by the Northern Ireland Fishermen's Federation (NIFF) and the North Irish Sea Fishermen's Association (NIFSA). The NFFO has taken advantage of this rivalry and now represents many of Northern Ireland's FOs.

The NFFO and SFF are similar in structure: each has an executive committee, comprising constituent representatives, with responsibility for formulating federation policy. Regional branches maintain regional coherence and provide a communication link for the rank-and-file members at grass-roots levels.

General policy and management decisions affecting the entire EU fleet are made by the Council of Ministers within the framework of the CFP. Within this management framework, member states are given a degree of autonomy for implementing regulations introduced by the Council: for example, the methods of reducing fishing capacity or distributing the national fish quota among the fleet varies between countries. In the UK, these powers rest with the Ministry of Agriculture, Fisheries and Food (MAFF), but decisions are made in consultation with the territorial departments for Scotland, Wales and Northern Ireland. There is no formal industry representation at this level. However, the national federations are consulted by the EU and UK fisheries departments on fisheries policy both through forward meetings and in writing. Regional and sectoral FOs are occasionally consulted over single-interest policies.

An annual meeting takes place between the UK industry and fisheries departments in early November. The industry is informed of the EU's intentions for the coming year, in particular the allocation of total allowable catches (TACs) for each stock – the UK gets a percentage set by the CFP which is the UK quota. At other times of the year, consultations with FOs may consider plans to introduce national or EU regulations as well as policies and regulations the industry would like to see introduced. For example, in 1993 the Ministry asked the industry to formulate a package of measures to address the problem of stock depletion and to meet EU fleet reduction obligations. Both the NFFO and SFF submitted very detailed proposals for a wide variety of options.

EFFECTIVENESS OF FOs IN UK FISHERIES POLICY UNTIL 1996

The participation of FOs in management and policy-making before 1996 was minimal, informal and of a crisis management nature. The annual meeting of industry and Ministry representatives in November only serves to inform the industry of the intentions of

the EU Council for the coming year; the industry has little prospect of influencing the Ministry's position at this late stage. This
kind of involvement, in relation to quotas, has been described by
the NFFO as 'cosmetic consultation' (Perry, 1995), i.e. they have
no real influence. Meetings sometimes occur when the government
is faced with a crisis, for example after the industry's widespread
condemnation in 1992 of the government's proposals to limit the
number of days British-registered fishing vessels might go to sea.
Coming together when there is a significant political problem such
as this has been described as 'crisis management' by the NFFO
(Perry, 1995).

There are two main reasons for such minimal industry involvement in decision-making: first, the UK's traditionally centralized
style of government and, second, the industry's lack of unity. These
factors provide the ideal conditions for 'divide and rule'.

UK's centralized management structure

In the days before the national federations were established, involvement of the fishing industry was virtually non-existent. Stocks
were managed primarily by technical measures (e.g. rules of fishing gear) and where appropriate through international access agreements. By the mid-1960s it was apparent that stocks could not support
this 'open access' approach and for the first time it became necessary to know the actual size of a stock and estimate what percentage could be taken as a tonnage (quota). This was the beginning
of a major intergovernmental battle with little room for industry
involvement other than the major players, e.g. the British Trawler
Federation (BTF) dominated by Aberdeen and Humberside vessels.
Following the declaration of 200-mile exclusive economic zones
(EEZs) by many countries during the mid and late 1970s, particularly Iceland, the UK distant-water trawler fleet and BTF collapsed
for lack of access to its resources. By 1977, the EEC declared its
sole competence for fisheries management and the UK government
began battling for the industry in Brussels, but with little coordinated advice from industry. Only since 1983, when the CFP was
agreed, has the industry developed a coordinated voice, but now
there is little room for manoeuvre and the government has got
used to working without the industry.

Disunity within the industry

Further development into formal involvement has undoubtedly been impeded by the lack of unity within the industry making it easier for government to 'divide and rule'. The existence of two national fishermen's federations reflects the different fishing culture and traditions either side of the Anglo-Scottish border. The SFF is dominated by trawlers and seiners operating in the northern part of the North Sea, landing demersal and pelagic fish. This is in contrast to the less uniform membership of the NFFO which represents members from all sectors: potters, set netters, pelagic, seiners and trawlers.

Although the two federations agree on many issues and their relationship is generally a good one with regular contact, significant differences exist at policy level. For example, the SSF agrees in principle with limiting days spent fishing as an effort control, coupled with more flexible quota management, whereas the NFFO vehemently opposes such measures (NFFO, 1993). Certainly, single-species quota restrictions can be disadvantageous for the Scottish trawler fleet prosecuting a mixed fishery where large quantities of more than one species may be caught. In addition, the often atrocious weather conditions in the north impose a natural limit to fishing time which makes this approach more appealing than rigid quota management.

Policy differences between the two main federations not only relate to national fishing patterns but also to competition from foreign fleets. The authorization of Iberian vessels into the Western Approaches in 1996 created so much antagonism in the south-west of England that it influenced NFFO policy to advocate the replacement of the current CFP with a system of interlocking coastal state management regimes (NFFO, 1996).

The SFF, on the other hand, have never advocated leaving the CFP, instead supporting reform from within. The threat of the EU member states' fleet, particularly those from Spain and Portugal, competing for their resources is less likely as both countries have no recent track records of fishing in the North Sea. Another reason for the SFF favouring continued CFP membership is the relatively high proportion of EU-agreed TACs for many demersal and pelagic stocks awarded to the Scottish fleet due to a lack of competition with other member states' fleets on their fishing grounds.

Although the government recognizes the two national federations as representing the UK industry, there are many fishermen and FOs not affiliated to either, principally for policy and financial reasons. For example, in 1994, some members of the SFF, particularly from the SWFPA, decided to support CFP withdrawal, leaving the Federation and forming a new organization called the Fishermen's Association Limited (FAL). Also four prominent west coast FOs (Western Isles FA, Highlands and Islands FA (static gear), Ullapool Boat Owners' Association and Orkney FA) have linked up to form a rival federation, the Federation of the Highlands and Islands Fishermen (FHIF) as they believe the SSF is too heavily influenced by the east coast trawlermen. Other FOs, particularly those confined to ports, comprising a high proportion of members exploiting non-quota species (such as lobsters and crabs) and therefore regulated more by local policies, prefer to concentrate their resources on local issues rather than make the financial commitment to federation membership. Poor representation thus undermines the two main federations' claim to represent the interests of all UK fishermen.

Also, although specialist committees within each federation provide for mainstream sectoral interests, constituent FOs are free to consult directly with government. They sometimes do so over issues of primary concern to themselves, such as quota management and technical measures as they feel there is a better chance of securing the best deal for their members by going it alone. Such single-interest representation clearly weakens the Federation's negotiating position.

Fragmentation of such a diverse industry in crisis is inevitable as each faction competes to survive. The Sea Fish Industry Authority's chairman recently stated 'one of the industry's biggest weaknesses is its fragmentation, this dulls political clout, because the Government is unable to get a collective view and is therefore hampered both in implementing new policies and in fighting on the industry's behalf' (Davey, 1996).

However, it could be argued that it is in the longer-term interest of the two national federations to remain separate given the EU's subsidiarity principle which encourages provincial representation (both the SFF and NFFO are members of the European Association of National Fishermen's Organizations, known as Europêche) and the prospect of a Scottish Parliament being established in the not too distant future.

Pressure-group tactics

In response to their lack of involvement, both federations frequently resort to pressure-group tactics which are intended to force decision-makers to take account of their concerns and incorporate their recommendations into fisheries policy. Such tactics have included: lobbying the EU and UK Parliaments through demonstrations; gaining the support of MEPs and MPs, particularly those with fishing constituencies; conducting fringe debates at party conferences; and taking militant action such as disrupting port trade, e.g. merchant shipping and ferry services. Fishermen can evoke a great deal of public sympathy and attract a considerable amount of media attention from a maritime nation steeped in fishing traditions and currently experiencing a wave of Euroscepticism.

One of the most effective forms of pressure-group activity undertaken by FOs was legal action by the NFFO against the UK government's days-at-sea legislation in 1993. The NFFO took the case to the European Court and although they lost, they still forced the government to abandon their plans to reduce fishing effort by preventing fishermen from working at certain periods of the year. This illustrates the industry's hidden powers against the government and it could be argued that FOs have the upper hand – the government not being as popular as fishermen.

HOW HAS INDUSTRY PARTICIPATION IN POLICY-MAKING CHANGED SINCE 1996?

The government has consulted the industry much more extensively than usual since 1996, over issues such as decommissioning, regional management strategies, technical measures and modes of dealing with the problem of flagships (foreign-owned boats allocated UK quotas). The Fisheries Minister announced at the end of January 1996 that MAFF was arranging a series of consultative meetings between the fishing industry and the EU on the next Multi-Annual Guidance Programme (MAGP) – a long-term strategy to reduce the EU fishing capacity – to run from the beginning of 1997 to the end of the century.

This unprecedented level of consultation during 1996 has even seen Prime Minister John Major and his senior policy advisers meet both federations. In a written reply to the SFF, the Prime Minister stated the government's commitment to improve communications with the catching sector: 'We have also been advocating a number of improvements, including better technical conservation, consultative committees to bring together fishermen and national EU fisheries managers, better liaison between the industry and our fisheries scientists' (Major, 1996). The Prime Minister also pledged the government's commitment to resolve the flagship problem in the Intergovernmental Conference (IGC). The SFF were immensely encouraged by the personal interest which the Prime Minister had taken in fisheries policy since meeting their leaders and hoped this contact would be maintained for the foreseeable future.

The government instigated the formation of the 'Fisheries Conservation Group' (FCG), comprising industry groups, fisheries scientists and nature conservation interests, which first met in February 1996. A central role of the Group is to develop the UK's input into EU-wide discussions by providing a unitary response, and in June 1996, the Group agreed technical-measure proposals requested by the EU. In response to these proposals, the Fisheries Minister said it was 'encouraging that the industry was able to agree on range of measures' and that 'such proposals will make a positive contribution to the wider review of fisheries conservation measures in the EU over the coming months' (*Fishing News*, 1996). Unfortunately, the new technical measures proposed by the EU in July 1996 appeared to completely ignore the recommendations put forward by the FCG.

The CFP Review Group, set up by the UK Fisheries Minister in 1995 to assess whether the industry's interests are well served by the CFP, is further evidence of the government's recent attempt to appease the industry. The Group's findings published in a report in July 1996 recommended that the interests of the fisheries sector are best served within the CFP, although they suggested a large number of improvements to it including greater regionalization and decentralization by the development of regional consultative committees for individual fishery zones. However, the NFFO's response to the report was that it 'ducks the central issue facing the fishing industry namely whether there should be a decentralized control or centralized management from Brussels' (Deas, 1996). Only one member of the Group represented the catching sector and the NFFO thought that 'the make-up of the Group was one of the main factors determining the report's output' (Deas, 1996).

WHY HAS THE GOVERNMENT ENCOURAGED GREATER INDUSTRY PARTICIPATION?

There are two interpretations of the government's recent drive for greater industry participation and supportive action. The first interpretation is that it is politically motivated, i.e. is straightforward electioneering, to appease Eurosceptic MPs (many of whom support UK withdrawal from the CFP), to avoid further aggravation in the run-up to the 1997 general election, and to win crucial seats in marginal fishing constituencies.

The other interpretation is that the government wishes to involve the industry in decisions in order to legitimize regulations and therefore improve the chances of compliance. For example, a well-publicized problem in recent years has been the total disregard of landing restrictions and the consequent establishment of markets for over-quota fishing (commonly known as 'black fish') that would otherwise be discarded. Unreliable catch and effort data undermine the government's scientists fish stock assessment work, which provides TAC advice to the European Commission and EU fisheries ministers. A leading MAFF scientist has been quoted as saying 'I believe the EU is moving towards a point where the fishing industry will take more of a role in stock assessment and development of more selective gears' (Pawson, 1996). In an attempt to obtain more accurate catch data, government scientists have started to liaise directly with fishermen in Scotland where the problem of black fish is particularly common (Perry, 1996).

Many fishermen take the view that the government's overtures to them are politically motivated. They question the sincerity of the government's aim to conserve fish stocks when it appoints ill-qualified fisheries ministers for short periods of time whose apparent objective is to avoid short-term political problems (such as reducing fishing pressure) and merely pretend to fishermen that they are trying to help them. Moreover, formal integration of the two main federations into the national decision-making process could deny them freedom to scrutinize policy and they may prefer to remain independent and retain the right to question the legitimacy of government action. Fishermen's organizations are disillusioned with the UK's governing strategy for fisheries and this dissuades them from forging formal links with what they consider an illegitimate governing regime. Furthermore, many fishermen share the federations' belief that despite greater participation in the national policy-making programme in 1996, they still have little influence over EU policy, which ultimately governs each member state's strategy. For example, CFP technical-measure revisions proposed by the Commission have largely ignored the recommendations put forward by the UK Fisheries Conservation Group. Also a controversial report (known as the 'Lassen Report') prepared by a group of fisheries experts for the EU and published in 1996 recommended a 40 per cent reduction in EU fishing effort to conserve fish stocks. The Commission translated this recommendation into a 40 per cent cut in the size of the EU fleet. The UK fishermen's federations rejected the Lassen Report not least because it had been drawn up without any input from the industry.

CONCLUSION

The approach to the 1997 general election and the government's small majority in Parliament during 1996 have certainly helped to politicize the plight of the fishing industry. The government has been forced into making a number of concessions, particularly to the two main federations who are able to exert greater pressure-group power. These include promises to seek to resolve the flagship problem, to review the CFP, and to consult the industry more extensively over policy decisions. As a result, the fishing industry is in a more powerful position and beginning to develop a more formal role in government advisory groups.

The government's efforts to help the industry are, of course, driven by the potential benefits they themselves would accrue by resolving the issues they have chosen to address. For example, solving the flagship issue would not only increase the government's electoral popularity in fishing constituencies but it could alleviate the government's problem of how best to reduce the UK's fishing capacity and avoid further penalization by the EU – the UK fishing industry is currently being denied EU restructuring grants due to overcapacity. Moreover, setting up working groups comprising industry representatives to decide national policies will reduce the government's workload, minimize industry criticism and increase the chances of compliance.

However, the government appears to be avoiding one of the central issues raised by the industry, namely the decentralization of management from Brussels and the formal involvement of the industry in policy-making and not merely advisory bodies. The risk of further upsetting our EU partners, particularly during a period dominated by the BSE beef crisis and the introduction of a single European currency, may well have deterred the government from addressing the question of decentralization. The unwillingness to tackle the question of devolving management responsibilities in the UK to the industry may also stem from the UK's historically centralized approach to governance, although there are cases of the catching sector being successfully integrated into management. For example, Sea Fisheries Committees (SFCs) are statutory regional management bodies that have jurisdiction out to six miles from baselines around England and Wales and their membership includes industry representatives. Also quota management has been devolved down to fish producer organizations, which were set up to promote marketing activities of members' vessels. On the other hand, the industry may favour informal negotiations as a means of being able to conduct behind-the-door deals, inevitable in such a diverse industry, without being held to account as would be the case in a more transparent and therefore legitimate arrangement.

Whether sincerely or for reasons of political expediency, the Conservative government has certainly made a concerted effort to appease the fishing industry by addressing some key issues that concern fishermen, including the incorporation of fishermen's representatives into policy advisory groups. In some respects, the industry has benefited from a closer relationship with government. For example, its advice has been taken on board by MAFF when formulating national policy. However, the industry does not appear to have come any closer to influencing EU proposals. Many hard-fought policy proposals agreed between industry and government have failed to materialize as the EU has chosen to largely ignore UK recommendations. The government has also failed to address the right of user-group participation in fisheries management, and whether it will succeed in obtaining a solution to the flagship problem at the IGC remains to be seen.

NOTES

1. I am grateful to Dr Stephen Lockwood and Barrie Deas for their comments on this chapter.
2. Following the declaration of 200 nautical mile exclusive economic zones (EEZs), the UK's distant-water fleet lost access to traditional fishing grounds (e.g. Icelandic waters), and consequently fishing pressure has increased closer to home.
3. Falling quotas due to declining fish stocks, increasing fishing effort, rising fuel costs and cheap fish imports have all contributed to a fall in profits for most fishermen.

BIBLIOGRAPHY

Anon. (1996) 'Technical Proposals in Preparation', *Fishing News*, 7 June, p. 2.
Davey, E. (1996) 'Matching Capacity to Resources', *Fishing News*, 26 July, p. 7.
Deas, B. (1996) 'Ducks Central Issue – NFFO', *Fishing News*, 2 August, p. 3.
Major, J. (1996) 'Safe with Us! – PM's Pledge to the Industry', *Fishing News*, 8 November, p. 1.
NFFO (1993) *Conservation: An Alternative Approach*, NFFO policy document for the EU (Grimsby: NFFO).
NFFO (1996) *Coastal State Management: An alternative to the Common Fisheries Policy*, NFFO policy statement (Grimsby: NFFO).
Pawson, M. G. (1996) 'Scientists Must Have Accurate Catch Data', *Fishing News*, 1 March, p. 7.
Perry, A. W. (1995) *Fish Stock Conservation and Management*, House of Lords Select Committee on Science and Technology, Sub-Committee 1 (Session 1994–95) HL Paper 80–1 (London: HMSO).
Perry, A. W. (1996) *Fish Stock Conservation and Management*, House of Lords Select Committee on Science and Technology, Sub-Committee 1 (Session 1995–96) HL Paper 25–1 (London: HMSO).

9 The Political Culture of Fisheries Management: an Anglo-Danish Comparison of User Participation[1]

Jeremy Phillipson

INTRODUCTION[2]

Differing political and institutional traditions in the United Kingdom and Denmark have played a significant part in the emergence of contrasting approaches to fisheries governance. While both systems remain arguably centralized in style, the Danish tendency is for a more formal involvement of user groups at the point of policy-making. In the UK, however, there have been significant inroads into the delegation of management functions at the regional or sectoral level. This chapter compares and contrasts the fishery policy systems of Denmark and the UK. This involves an analysis of the institutional approaches to policy formulation and implementation and in particular the extent to which fishermen's organizations participate. The chapter considers whether either state can learn from the other's institutional experience.

Fishermen's organizations (FOs) in most of the major fishing industries of the European Union have been assimilated within systems of policy formulation and implementation. The aim has been to draw upon the knowledge base and skills of the fishing industry to provide more sensitive and practical approaches to governing fisheries, and thereby to facilitate conflict resolution, the legitimization of regulatory systems and more integrated management approaches. The assimilation process has, however, been significantly uneven in extent and contrasting in form for different national systems. This chapter compares approaches to industry participation in Denmark and the United Kingdom, which vary significantly in the formality and level of participation and in the location of user involvement within the policy system.

Approaches to participation within policy processes are partly a reflection of the political culture of fisheries management. This comprises the political ethos and institutional tendencies within which fisheries management mechanisms and approaches are framed. It is difficult to divorce fisheries from the wider national political culture which incorporates the democratic or participation tradition and which is influenced by organizational and management history within the broader policy environment. In some cases there are 'participant', in other cases more bureaucratic or 'subject' national political cultures (Almond and Verba, 1963). At the same time, however, there is variance concerning the degree of participation within nations in different industrial sectors or subsectors. This also fluctuates over time.

Diversity in institutional approach between the UK and Denmark, arising from their contrasting national political cultures together with the institutional development of their individual fishing sectors, is in fact enabled by the particular division of responsibility within the European Community framework and the Common Fisheries Policy (CFP). While the European Union is responsible for setting the frame conditions within which both the Danish and UK fishing fleets operate, aspects of policy implementation are primarily the responsibility of individual member states. Hence, while both nations generally utilize a similar set of management tools (quotas, licensing, technical measures, etc.) there is scope for considerable diversity in terms of the detailed arrangements and institutional mechanisms for applying and generating such tools and for achieving particular objectives (such as enforcement, fleet restructuring, etc.).

This chapter does not review the detailed aspects of management tools existing in the case countries, nor does it analyse directly the function and effectiveness of fishermen's organizations. The key focus is upon institutional approach and participation. Attention first turns to the systems of policy formulation (the process by which policy decisions are generated) and subsequently considers approaches to implementation (the means of carrying out such decisions). Finally, some central issues concerning user participation are considered.

POLICY FORMULATION AND USER PARTICIPATION

Policy formulation in Denmark – 'Scandinavian Negotiation Economy'

Denmark has a well established tradition of interest participation which dates back to the cooperative farming movement in the early nineteenth century. Most industrial sectors, including fishing, have been assimilated within the policy system in order to resolve conflicts, influence policy development and sector objectives, and implement management strategies. Emphasis is placed upon giving all groups with an interest a say in the decision-making process (Nielsen, Vedsmand and Friis, 1995). The Danish situation is indicative of the broader picture in Northern Europe that has been termed the 'Scandinavian Negotiation Economy'. Nielsen and Pedersen (1988: 82) define such an economy as an

> organizational structuring of society, where an essential part of the allocation of resources is conducted through institutionalized negotiations between independent decision-making centres in the state, organizations (employer/employee) and corporations ... the decision-making process is conducted via institutionalized negotiations between the relevant interested agents, who reach binding decisions based on discursive, political and moral imperatives.

Castles (1967: 62) ascribed the emergence of this unique approach to a homogeneous political culture, an emerging consensus about the role of state, party and interest organization, and to the evenness by which the process of industrialization took place in Scandinavia:

> The lack of political alienation in Scandinavian society has meant there have been no fears that one class or group would arrogate power and become 'a managing committee' to exploit the others, and consequently there has been little distrust of large centrally organized interest organizations. Indeed, it is possible in many ways to talk of Scandinavia as the totally organized society'.
>
> (p. 65)

Institutionalization of users, however, has not reached the highly developed levels in fisheries as it has in other sectors of Denmark (Vedsmand, Nielsen and Friis, 1995). Some ascribe this to the lower significance of the fishing industry in terms of GNP. It may, however, be that the industry is simply a late entrant into the policy arena; it was not until there was increased regulation of the fisheries sector during the 1970s that there was a need for more user participation and conflict resolution. There are also many current difficulties specific to fisheries which may block more significant assimilation, notably the lack of consensus and an acute crisis.

Nevertheless, throughout most of the century and particularly since the late 1970s, Danish fishermen's organizations have been formally integrated within the policy formulation process and have been able to influence the principles and objectives of resource allocation and regulation before the formulation of policy (Vedsmand, Nielsen and Friis, 1995). Currently, integration occurs primarily through two advisory boards which advise the Ministry of Agriculture and Fisheries. The Regulation Advisory Board and the EU Advisory Board both consist of a wide range of interests and sectors, although the catching sector has the greater proportion of seats on both (Table 9.1). Of key importance to the system of Danish fisheries regulation is the Regulation Advisory Board which meets on a monthly and needs-determined basis. Central to its remit is to provide advice concerning the seasonal and sectoral allocation of quotas. The Board incorporates several working groups and subcommittees where, again, the catching sector is in ascendancy. They consider a range of issues including existing and alternative regulatory systems, structural measures or issues facing particular fisheries. The EU Advisory Board advises the Minister on Commission proposals and on issues prior to meetings of the European Council of Ministers. Advice is also relayed to the European Board within the Danish Parliament.

Overall, through both advisory systems, the catching sector is able to exert considerable influence on structural and regulatory policy within Danish fisheries, particularly in relation to quota allocation. The yearly advice from the Danish Fishermen's Association is often synonymous with decisions of the Regulation Advisory Board and the Ministry. Advisory success on quota issues is partly attributable to two key factors: the level of organization within Denmark in terms of the number and coverage of fishermen's organizations and the associated system of internal representation. The Danish fishing industry is organized and centralized through one representative organization, the Danish Fishermen's Association, *Danmarks Fiskeriforening* (DF), which comprises over 80 per cent of Danish fishermen and vessel owners and 86 constituent local fishermen's

Table 9.1   Constituents of the Advisory Boards

	Regulation Advisory Board	EU Advisory Board
Danish Fishermen's Association	2	4
Producers' organizations	1	1
Association of Danish Fish Processing Industries and Exporters	1	3
Association of Fish Meal and Fish Oil Manufacturers in Denmark	1	1
Association of Canned Fish Manufacturers in Denmark	1	1
Association of Trout Exporters in Denmark	—	1
General Workers Union in Denmark	2	1
Consumer Council	—	1
Danish Institute for Fisheries and Marine Research	1	—
Ministry of Agriculture and Fisheries	1	2
Total	10	15

organizations. It was formed following a merger in 1994 of two former associations, the Danish Fisheries Association (representing the inshore industries of Zealand, the islands on the east coast and the inshore fishermen of Bornholm and Jutland) and the Seafisheries Association of Denmark (with members from West Denmark and North Jutland). The merger attempted to strengthen the industry's voice within the advisory arena by providing a more united view and outlook. DF has also been able to manage the fragile balance of internal regional interests, primarily between the larger and more numerous vessels in the North and West and the inshore vessels in the East. While membership of the Central Board of DF reflects regional membership numbers, the distribution of votes is weighted on the basis of regional interest. This is complemented by Regional Fishing Area Committees within DF which advise upon quota allocation within specific waters.

The advisory arrangements outlined above are criticized by some within the catching sector for not going far enough (Nielsen, 1994). Part of the reason for this may be that these arrangements are relatively undeveloped or embryonic when compared to the Danish context of participation. For example, the remit of the advisory arenas does not extend to key areas of policy; FOs have less influence upon the overall objectives for the sector and more particularly upon the fixing of total allowable catches (TACs) and quotas at a European level. Where they do have a remit, it is only advisory and non-executive. There have also been a number of other areas of criticism. Given the reduced size of fishing quotas, some sense that the industry has been burdened with a thankless task in quota distribution. Moreover, the delicate balancing of regional interests involved is criticized as contrary to the needs of long-term strategic management. Finally, outside the context of subcommittees and working groups, which appear to function effectively given more focused representation and objectives, some have noted a paralysis of decision-making within the wider advisory boards faced with severe problems in achieving agreement across the wide range of interests involved, although this has been partially negated in recent times following the FO merger.

Policy formulation in the UK: 'subject' political culture

Criticisms of the Danish system of centralized consultation, on grounds that participation in fisheries is undeveloped or embryonic when compared to other sectors, are placed more in perspective when one considers the extent of user participation within the UK. Traditionally, FOs have been relatively peripheralized from the policy process, involved in lobbying rather than negotiation and reactive rather than proactive in approach; they have been subjected to policy decisions without any real say. This reflects the broader tendency within the UK for centralized and more bureaucratic systems of governance:

> ... the British state is highly secretive, with access to central government controlled by civil servants. As a result the British state has a tendency towards elitism. Decisions are made by a small number of Ministers and civil servants.   (Smith, 1993: 9)

> Consultation does not take place within an arena that encourages a consensually-established rationality ... It is a style which reserves to government the right to decree authoritatively what is in the 'national interest' ... the dominant tendency is for interests to be taken into government rather than the latter moving towards society.   (Smith, 1989: 238)

This is not to say that FOs are not consulted; in fact, it could be argued, that the level of consultation has increased in the mid-1990s. Fishermen's organizations and a range of other groups are regularly asked to comment on consultation papers on policy and regulation matters. There are frequent meetings with some FOs on matters of policy implementation (as noted later in the chapter). The most integral form of consultation is informal and involves meetings between government departments and their respective national representative federations, the National Federation of Fishermen's Organizations (NFFO) based in England and Wales, whose main lobbying target is the Ministry of Agriculture Fisheries and Food in London (MAFF), and the Scottish Fishermen's Federation (SFF) whose target is the Scottish Office Agriculture, Environment and Fisheries Department in Edinburgh (SOAEFD). Ministers repeatedly point out to the media and the fishing industry that they are willing to consult and that they are in regular contact with the catching sector.

Three recent *ad hoc* developments have also been government initiated. They include the CFP Review Group set up in January 1995, which, as the name suggests, aimed to assess the CFP and alternative industry-generated strategies, and which in July 1996 unveiled its final recommendations for reform (CFPRG, 1996). Late in 1995 the government also initiated a Fisheries Conservation Group comprising scientists, civil servants and industry, which intended to develop technical conservation measure proposals and encourage greater understanding between industry and scientists through a series of meetings. Finally, the government has recently proposed to the European Union the establishment of regional consultative committees which will bring together industry and government on a 'regional seas' basis (e.g. North Sea, Baltic Sea, Mediterranean, etc.) at a European level (Hansard, 1996). Such a proposal is also contained within the final report of the CFP Review Group.

The system of UK user participation contrasts with the situation in Denmark in several ways. Of immediate significance is that representations in the UK are primarily single interest and not made in conjunction with other sectors or interests. Crucially the system is less transparent and public than in Denmark, with key discussions occurring often informally and behind closed doors. Such a system does not create the impression for many in the industry that effective or timely consultation takes place. The problem is not, however, simply one of perception. There is considerable agreement that industry advice is not integral to the policy process. Blame, however, is more often than not placed with the industry itself rather than with a bureaucratic decision-making system. The lack of industry unity and the sheer number of representative factions within the catching sector, each to a certain degree competing for the ear of government, limit the possibility of more formal arrangements for negotiation. A further problem is that the current national federations, the SFF and NFFO, are not truly representative. Numerous fishermen's associations remain outside these federations, for policy, financial or even personality reasons. Some within are sometimes critical of power imbalances and decision-making cliques. Other industry groups, such as the Save Britain's Fish campaign (SBF), with objectives far removed from mainstream political opinion (in this case withdrawal from the CFP), appear to lack the credibility to be effectively incorporated in the policy arena. They must resort to high-profile public campaigning to achieve their objectives.

However, as noted earlier, with reference to recent consultative developments, it is possible that the UK approach to user participation within policy formulation is undergoing change. The increasing crisis facing the fishing sector, notably a crisis of confidence in the regulatory system, has stimulated a sense of urgency for greater industry participation. It is, however, premature to assess the significance of these developments and to ascertain whether they represent a significant incorporation of fishing interests. So far, they do not appear to shift the approach to one of effective partnership between government and industry characteristic of the situation in Denmark. Most of the developments are clearly *ad hoc* and some have faced severe criticism, notably the CFP review group for its representation which comprised only one representative from the catching sector, two academics, and one person from each of processing and marine environment interests. It may be that these high-profile developments are intended simply to help the legitimation cause and that they have little real policy impact.

POLICY IMPLEMENTATION AND USER PARTICIPATION

While the UK fishing industry remains comparatively peripheral to the policy formulation process, it has become closely involved at the stage of policy implementation through the sectoral quota management system and inshore fisheries management. By contrast,

Danish FOs are clearly less active at the sectoral or regional level and their prime activity remains geared towards the central advisory arenas.

Producer organizations

Producer organizations (POs) were established as marketing organizations to ensure that fishing is carried out along rational lines and to improve the conditions of sale for members catches. Their organizational form is relatively similar in both Denmark and the UK and throughout the EU. The notable exception is that in some member states like the UK, POs also undertake quota management responsibilities.

UK quota management is partially administered by producers' organizations under the *sectoral quota management system* and is coordinated jointly by MAFF, SOAEFD, and the Department of Agriculture for Northern Ireland (DANI). The system covers all demersal quota stocks and three pelagic stocks. At the start of the fishing year, each demersal species quota is divided for management purposes into allocations to three vessel groups: those vessels of over 10 m belonging to a PO, those of over 10 m not belonging to a PO and the 10 m and under fleet. These final two categories form the so called non-sector, which is managed by the fishery departments with regular consultation with industry representatives. For the major pelagic fisheries, the quotas are divided into sectoral allocations (to PO vessels), a small non-sector allocation, and individual allocations granted to certain purse-seine and freezer vessels. Allocations to POs are on the basis of the aggregated track records of over 10 m member vessels using various reference periods and each PO is able to internally distribute the allocation to its members in the manner which it sees fit. The PO quota management role necessitates regular contact with the relevant government department who monitor the PO catch reports and internal mechanisms. The POs also meet annually with the Minister in advance of the annual TAC negotiations. There appear to be a number of emerging problems concerning POs in the UK, notably their spiralling number (currently 19), a lack of geographical coherence, doubts over their ability to impose in-house discipline and a suboptimal engagement in marketing activities (Symes et al., 1995). For some of these reasons the role and structure of UK POs is currently under review (MAFF, 1996).

In marked contrast to the UK there are only three POs in Denmark, the Danish Fishermen's PO (*Danske Fiskeres Producent organisation*) comprising 90 per cent of Danish vessels from across Denmark, the Purse Seines PO (*Notfiskernes Producentorganisation*) incorporating 11 purse seines within the Danish pelagic fisheries and the Skagen Fishermen's PO (*Skagenfiskernes Producentorganisation*) with membership concentrated primarily in the base port in Skagen, North Jutland (Vedsmand, Nielsen and Friis, 1995). Danish POs are primarily occupied with marketing issues involving quality control and the implementation of EU minimum withdrawal prices. They have been reluctant to become more involved in quota management issues and are subsequently not as central within the management framework as in the UK. Only the Purse Seines PO has managed herring and mackerel quotas for its members. Partially, the reason for a lack of quota responsibility has been PO criticism of the means of quota allocation within the Regulation Advisory Board, which they feel does not take sufficient account of market considerations. In the past it has also been in the interests of the leading representative organizations for POs not to become too economically powerful through an expanded remit. They have favoured the status quo in terms of role boundaries, seeing POs as organizations with a marketing function and categorizing fishermen's associations as groups with regulation interests. In the UK this division has become increasingly blurred through a politicization of POs, and more specifically the incorporation of POs within the executive committee of the NFFO in England and Wales.

Regional participation

Both the UK and Denmark display broadly centralist systems of government, with decision-making power concentrated in Whitehall and the *Folketing*. In both instances, however, the central state is complemented by a regional administrative apparatus which, to a greater extent in the UK, involves and influences approaches to user participation at the regional level in fisheries.

First, there is the UK provincial state system which has special significance for participation as well as the fisheries management system. Here, fisheries are managed through cooperation between the provincial departments of MAFF in England and Wales, SOAEFD, DANI and the semi-autonomous island assemblies for the Channel Islands and Isle of Man, although within these arrangements MAFF maintains the lead department status. This approach provides several potential access points to the government for the regional industries, although some might argue that it further distances the provincial industry from the effective power base within MAFF.

Second, both Denmark and the UK have tiers of local government which operate within the wider parameters of the centre. In England and Wales, however, it provides users with a direct input into inshore fisheries governance via the regionally based Sea Fisheries Committees (SFCs) which straddle local authority and fishing industry interests. The 12 SFCs have responsibilities for management, enforcement and enhancement within the six-mile limit. Half of the seats are allocated to constituent county councils as representing the financial backbone to the Committees. One seat is allocated to a representative from the Environment Agency and the remainder to those appointed by MAFF. These are persons acquainted with the needs and opinions of fishing interests within the Committee's district and are chosen with advice from local fishermen's organizations and district inspectors. They include both active fishermen and shore-based individuals (agents, processors, environmentalists, retired fishermen). Each SFC is provided with a single vote in the Association of Sea Fisheries Committees which also meets annually with the Minister. SFCs, and more specifically local industry participation, face a number of challenges, notably a realignment of existing functions to include powers to introduce by-laws regulating fisheries for marine environmental management purposes, and an insecure financial base, since local government funding is not statutorily safe. Their powers are also restricted in a number of ways, notably in terms of the protracted process of local by-law formation.

In Scotland and Northern Ireland the absence of SFCs, given alternative provincial approaches to inshore governance and participation, poses a considerable challenge to the integrity of inshore fisheries management. Local participation involves primarily voluntary actions and occasional local management forums associated with conflict resolution. In addition, some Scottish Regional Councils encourage initiatives for fishery-dependent regions in conjunction with FOs. This kind of activity is emulated in dependent regions in Denmark, notably in Bornholm and North Jutland. Both have seen integrated regional initiatives to develop their fishing industries comprising a multiplicity of interests from local government, the catching sector and processing interests.

In Denmark there are also a limited number of regionally specific examples of user participation in regulation activities. These
include the Matjes Committee, consisting of regional FOs that
manage the Matjes herring fishery in the North Sea and Skagerrak.
There is also an experimental 'days at sea' system within the Kattegat
for the sole and nephrops fishery, developed jointly by DF and the
Ministry. As a whole, however, in Denmark there has not been a
comprehensive or equivalent devolvement of management responsibilities to a local institutional framework like that seen in England and Wales. Local Danish fishermen must seek representation
through the structures of the DF and, ultimately, the advisory boards
to influence local policy.

ISSUES AND DILEMMAS OF PARTICIPATION

The development of user participation within systems of policy
formulation and implementation has clearly taken different paths
in Denmark and the United Kingdom. Both case studies raise important issues of relevance to participation systems in general and
to the specific cases. The UK does not have an open and formal
system of consultation such as that seen in Denmark. If there has
been an attempt within the UK to legitimize the policy system through
consultation processes this has not been an obvious success. At the
same time, the Danish may learn from the UK's experience of
delegation to producers' organizations and Sea Fisheries Committees who are responsible for a range of centrally determined administrative tasks. Delegation in these instances may offer more
cost-efficient, practical and regionally responsive management approaches. It may also help to distance central government from
any criticism arising from policy failure or the politically unpopu-
lar decisions associated with day-to-day fisheries management.

It is, however, difficult to prescribe solutions from one state to
the other as both of their industries are undoubtedly conditioned
by their own political contexts, institutional backgrounds and norms
of participation. User participation in both the UK (within the system of policy formulation) and Denmark (in terms of policy implementation) is curtailed by a range of factors such as aspirations
within both industry (unwillingness to accept responsibility) and
state (unwillingness to concede responsibility). The industry can
suffer from apathy or entrenchment of attitudes and roles and there
is sometimes the fear that they will have reduced freedom if formally

assimilated within the policy process. Those who are already posi-
tioned favourably within the institutional structure are eager to protect
the status quo. There are also organizational problems, such as
the lack of unity or doubt over the representative capacities of
FOs and advisory arenas. Finally there are capability issues relating to levels of resources and the professional credibility of FOs
within the policy system.

Several of these factors can help explain the reluctance or inability of the UK government to effectively incorporate the industry within the policy formulation process. Here there is also a question
of objectives: assuming that the government has clear objectives
for the fishing industry (some would question even this), it appears
that for whatever reason it intends to fulfil them without fully har-
nessing user cooperation at the decision-making stage; the choice
of a 'go it alone policy' is ineffectively opposed by a structurally
weak industry. At the same time, government objectives and approaches are often significantly different to those of the fishing
industry and there are different interpretations of appropriate crisis response. The government's objectives rely heavily upon the
scientific basis of management in terms of stock assessment and an
emphasis upon downsizing of fleet through market forces. Fundamental conflicts that would be likely to arise under conditions of
partnership are therefore avoided through separation of industry
and state. It is unlikely, however, that longer-term fisheries management will succeed in the UK without industry cooperation, given
the hiatus of legitimacy and knowledge within the present system.
Current attempts to provide these benefits through the allocation
of responsibilities to POs and SFCs do not suffice, given the
peripheralization of the industry from central decision-making and
the limited extent of the delegated tasks.

The two cases also highlight a range of key dilemmas relating to
user participation systems. Indeed, many of the criticisms of existing consultative arrangements are often reducible to one or more
of these paradoxes. Firstly, there is a *dilemma of representation* which
involves questions concerning the nature, and more particularly the
breadth, of representation, whether systems should incorporate a
multiplicity of interests, as is the tendency in Denmark, or a more
single-interest basis as in the UK. Both approaches have their advantages. On the one hand, under the plural system, there is the
possibility of more integrated approaches to fisheries, which are
increasingly important given the challenges facing the industry. On

the other hand, single-interest structures mean that users have increased scope to handle executive responsibilities given more focused objectives and demands. The Danish industry is often critical that it is restricted to an advisory role, but the multiplicity of the representation systems involved would seem to prevent anything else. The need to incorporate other interests from outside the fishing industry, such as environmentalists, is a particular concern for the UK catching sector which is anxious to prevent infiltration of the existing fisheries policy arena; its lack of unity may, however, mean it is less in a position to resist.

Linked to the dilemma of representation is that of *transparency*. For maximum legitimacy benefits, consultation systems should be open, public and transparent. Membership of FOs can then see that consultation is taking place. However, transparency may slow down and complicate the decision-making process. The dilemmas of representation and transparency are thus interrelated; if the 'policy community' is a relatively closed shop and of a single interest, this might lead to more efficient and simple decision-making approaches, but to the detriment of more integrated, open, democratic and even legitimate management systems. The legitimacy of regulatory decisions may also wane with multiple systems of representation where the interests of FOs are subsumed with those of other groups. The question comes down to a choice of objectives: whether one is aiming for a legitimate system (which would suggest an open approach with single catching sector representation), an efficient decision-making process (where closed, single-interest systems are in order), or where importance is placed upon democracy and integrated approaches (favouring multiple interest and open systems) such as the situation in Denmark. A partial compromise has been found in the Danish case which has utilized working groups and subcommittees: these are closed bodies with less breadth of representation, but they are set within the more integrated, multiple and transparent contexts of an advisory board or regional initiative.

Crucial to the question of user participation within systems of policy formulation and implementation is the *dilemma of crisis*. As the problems of the fishing industry intensify there is an increasing need to engender cooperation, draw upon user knowledge and legitimize the management system. It is difficult, however, to predict positive institutional changes within such an environment. Under conditions of crisis the fishing industry appears to fracture and further politicize, which poses obvious problems for user involvement. From another perspective there may be unwillingness to become involved in the management of a system undermined by crisis and decline. The criticism that the DF is simply involved in the distribution of poverty within the Regulation Advisory Board is relevant here. As a whole it is difficult to identify whether user participation is, in fact, more likely to emerge within fair weather or crisis environments. Legitimacy and user knowledge are arguably essential in both contexts in order to avoid and ameliorate crisis. However, if users are not incorporated as a matter of routine (there may appear less need to do so if an industry is in a healthy state and this in itself may be a contributing factor to crisis), then it is perhaps more difficult to incorporate them after a crisis situation has emerged. In the UK it may be that recent tendencies towards user participation are in response to the encroaching problems, but there are obvious difficulties of assimilation under such conditions and the problem is unlikely to ease unless there is an immediate loosening of the overarching resource and structural angst.

It is possible that the industry will be able to cooperate within a user participant system if it is given a real forum in which to do so (as in Denmark where a well developed advisory system encouraged greater unity among FOs) as well as integral or executive responsibility within the policy system. Here, however, there is a significant challenge in formulating the correct balance of influence. The fishing industry cannot be given too much influence in key areas of fisheries policy, but its complete exclusion from vital decisions may cause significant legitimacy problems. This *dilemma of influence* is very relevant to the UK and Denmark: as members of the European Union, both industries are one step removed from important aspects of decision-making in Brussels. It is reasonable to suggest that it is at the European level where maximum benefits of user participation might be gained, but there is criticism that TAC and quota arrangements are framed there without effective industry consultation. Fishermen's organizations cannot be given executive responsibilities in these central and political areas of fisheries policy but their advice needs to be taken effectively and openly into consideration. Currently FOs are represented through the European Association of National Fishermen's Organizations (Europêche) which holds 18 of the 45 seats within the Advisory Committee on Fisheries together with representation within its market, structure and resource subcommittees. The overall committee meets relatively infrequently and, some would suggest, after key policy decisions have been tabled. Europêche also lobbies informally across the range of EU institutions, notably the Commission and the European Parliament. It is, however, relatively constrained, not only by the formidable task it faces in presenting a united viewpoint of the EU catching sector, but also in terms of its administrative resources. Overall, it can be concluded that effective systems of user participation at the national level must be nested within European approaches. Recent proposals for regional consultative committees, as noted earlier in this chapter may signify a step in this direction.

CONCLUSION

By examining a range of factors relating to institutional background and democratic tradition, there is an improved understanding of the contrast between approaches to user participation in fisheries in Denmark and the United Kingdom. At the level of policy formulation there is clearly much more formal involvement of FOs in Denmark than in the UK. This is based upon a more highly organized structuring of fishermen's organizations and a favourable political culture in Denmark. However, there have been significant developments towards devolved management in the UK at the level of implementation involving regional and sectoral institutions, a trend which does not feature strongly in the Danish case. These differences highlight the diversity of policy systems in the two states; they confirm that particular political and institutional settings produce diverse solutions and approaches.

At a basic level one can assume that optimum government – industry relations in fisheries at times of both crisis and success are characterized by dependency and, therefore, mutuality. In both the UK and Denmark, however, there are barriers to a full integration of users, and some of these barriers, notably the politicization of the industry, may be intensified during times of crisis. This is just one of a complex of dilemmas relating to the development and architecture of participation systems.

In practice it is very difficult to assess participation systems, to ascertain the level of influence held by FOs and to decide whether incorporation actually enhances policy outcome. This is particularly the case if the industry faces acute difficulties when the basis of the most beneficial of management structures may be challenged.

Policy outcomes are the result of a combination of influences from within and outside fisheries. Furthermore, evidence of consultation or incorporation does not necessarily mean influence or that participation is integral. That an outcome is synonymous with industry advice may be a consequence of these other influences rather than the advice itself; industry opinion may simply mirror state intention, and consultation in these instances may only be an exercise to legitimate a particular state aim or approach to governance arrived at by other means. In states with a long-established habit of acknowledging the interests of users one may be more prone to assume that consultation exercises are integral to the decision-making fabric.

NOTES

1. The author wishes to thank Tomas Vedsmand of the Research Centre of Bornholm, Nexø, Denmark, for his valuable comments and criticisms during preparation of this chapter.
2. Some of the issues referred to in this chapter were developed as part of the EU-funded project on Devolved and Regional Management Systems for Fisheries (AIR-CT93-1392, DGXIV SSMA) based at the University of Hull.

BIBLIOGRAPHY

Almond, G. A. and Verba, S. (1963) *The Civic Culture: Political Attitudes and Democracy in Five Nations* (Princeton, NJ: Princeton University Press).

Castles, F. G. (1967) *Pressure Groups and Political Culture: A Comparative Study* (London: Routledge & Kegan Paul).

CFPRG (1996) *A Review of the Common Fisheries Policy Prepared for UK Fisheries Ministers by the CFP Review Group, Volume 1: Conclusions and Recommendations* (London: MAFF).

Hansard (1996) No. 1720, 1–3 April.

MAFF (1996) UK Fisheries Departments: Consultation Paper on Fish Producers' Organizations (London: MAFF).

Nielsen, J. R. (1994) 'Participation in Fishery Management Policy Making: National and EC Regulation of Danish Fishermen', *Marine Policy*, 18: 29–40.

Nielsen, J. R., Vedsmand, T. and Friis, P. (1995) *Devolved Fisheries Management Systems: A Discussion on Implementation of Alternative Fisheries Co-Management Models in Denmark*, Working Paper, North Atlantic Regional Studies, Roskilde University.

Nielsen, K. and Pedersen, O. K. (1988) 'The Negotiated Economy: Ideal and History', *Scandinavian Political Studies*, 11, 2: 79–101.

Smith, G. (1989) *Politics in Western Europe – A Comparative Analysis*, 5th edn (Guildford: Biddles Ltd).

Smith, M. J. (1993) *Pressure, Power and Policy: State Autonomy and Policy Networks in Britain and the United States* (New York: Harvester Wheatsheaf).

Symes, D., Crean, K., Phillipson, J. and Mohan, M. (1995) *Alternative Management Systems for the UK Fishing Industry*, EU-funded Research Project (AIR-2CT93-1392: DG XIV SSMA) Devolved and Regional Management Systems for Fisheries, Working Paper 5, School of Geography and Earth Resources, University of Hull.

Vedsmand, T., Nielsen, J. R. and Friis, P. (1995) *Decision-making Processes in Danish Fisheries Management: Capabilities and Aspirations of Danish Fishermen's Organizations*, Working Paper, North Atlantic Regional Studies, Roskilde University.

10 Sectoral Quota Management: Fisheries Management by Fish Producer Organizations
John Goodlad

INTRODUCTION

This chapter describes how the management of UK fisheries has developed into a system of sectoral quota management (SQM) over the past decade. The advantages and disadvantages of this particular management system are analysed. It is suggested that the future development of the SQM system could result in a system of individual transferable quotas (ITQs). Alternatively, it could also develop in the opposite direction, in terms of communal ownership of fish quotas by fish producer organizations (POs).

The example of Shetland is used throughout the chapter. Shetland is now becoming one of Britain's principal fishing centres and has been at the forefront of many fisheries management initiatives in recent years. In particular, the Shetland PO (SFPO) has played a crucial role in the establishment and subsequent development of the SQM system.

THE ORIGINS OF SECTORAL QUOTA MANAGEMENT

Before 1984, fisheries within the UK were managed on the basis of vessel quotas which were set by fisheries departments. These were usually set on a fortnightly or monthly basis. In other words, if North Sea whiting or West of Scotland herring were subject to catch limits, all UK fishing vessels would receive the same fortnightly or monthly quota from fisheries departments. From time to time these quota allocations varied depending on vessel size. Such a system took no account of regional variations nor of the requirements of different sectors of the UK fleet. This system was also rather remote in that fishermen, through their organizations, were not directly involved in the decision-making process as such. There were regular consultations between the industry and fisheries departments, but the final decisions on setting vessel quota limits were made by government and not industry.

The absence of any real involvement in the decision-making process, together with the lack of a regional or sectoral dimension in the quota allocation process, led to much criticism. Nowhere was this criticism more marked than in Shetland. During the early 1980s there was a lucrative industrial fishery for sand eels around Shetland during the summer months. This fishery regularly attracted a large number of vessels that would otherwise have been catching white fish. A small number of white fish trawlers were left to supply the local white fish processing plants. In 1983 the UK haddock quota limits were particularly poor during the summer months. One unforeseen result of this quota was that the limited number of Shetland white fish vessels were unable to land enough haddock to supply the requirements of the local fish processing industry. By the time the sand eel fishery had finished in September, the haddock catch limits had been raised. But, although the entire Shetland white fish fleet was now able to fish for haddock, the 'summer haddock fishery' had been lost. The possibility of landing sufficient haddock during the summer months, when a large proportion of the fleet usually diverted to industrial fishing, only seemed possible if larger per vessel quotas could be allocated, something which was patently impossible under quota system which existed at this time.

In view of this, the Shetland fish catching and fish processing industries argued that a more flexible system of quota management was necessary in order to take account of the particular circumstances pertaining in Shetland at this time. The SFPO had been established in December 1982 in order to try and improve the marketing of its members' catches. Since the effective marketing of the haddock catch to the local fish processing industry was being prevented by an inflexible quota management system, the SFPO began to promote an alternative system.

Why not, it was suggested, allocate to the SFPO that share of the UK haddock quota which the Shetland fleet would normally catch during a full year? This quota could then be shared between member vessels in the manner best suited to local market conditions. The Scottish Office Fisheries Department was persuaded and the SFPO received a haddock quota allocation for 1984. The first tentative steps had been taken towards the SQM system.

This experiment, in so far as the Shetland fishing industry was concerned, was successful. Unlike the previous system, the SQM system allowed fishermen, through their POs, to manage fisheries. At the same time fisheries departments could effectively devolve the increasingly burdensome and time-consuming job of fisheries management on an individual vessel basis. As a result there was a rapid move towards adopting SQM throughout the fishing industry at large. By 1985 most Scottish POs and a few English POs were managing their haddock, cod, whiting and saithe fisheries under the SQM system. Within two years the herring and mackerel fisheries were also being managed under this system.

THE PO SYSTEM

POs are a relatively new type of fishermen's organization. Unlike the long-established trade organizations (many of which have been established for most of this century), POs are the direct result of British membership of the EU. The first British PO was established in 1973 – the year Britain joined the then Common Market.

Under the terms of the EU Common Fisheries Policy (CFP), POs play a central role in the common organization of the market. The principal objectives of the POs throughout Europe are 'to encourage rational fishing and to improve conditions for sale of their members' products'. In order to achieve these objectives all European POs have a responsibility to implement the marketing regulations of the CFP. The PO system enables fishermen to enjoy the benefits of the EU minimum price scheme and market support mechanisms. POs therefore must ensure that fish landed by member vessels are properly graded according to EU size and freshness criteria. The EU official withdrawal price (i.e. the minimum price below which fish cannot be sold) must be strictly observed if PO member vessels are to benefit from the market intervention system (i.e. the system whereby financial compensation is paid for fish which cannot be sold at the official withdrawal price).

All POs throughout Europe are now involved, to a greater or lesser extent, in the implementation and administration of the EU marketing regulations. Some POs have become involved in related activities such as the establishment of quality control systems, the marketing of fish and the establishment of fish processing plants. It is only within the UK, however, that POs have come to play a central role in fisheries management. This new role for POs was recognized in 1993 when the EU marketing regulation was amended to allow POs, at the discretion of member states, to manage national catch quotas. With this change in the relevant regulation, the EU has clearly signalled its approval of fisheries management by the PO sector. It will therefore be interesting to see if other member states follow the UK approach and develop fisheries management systems based on PO participation.

Within the UK there is now a total of 19 POs. These are largely, although not entirely, regionally based. These POs now represent the vast majority of Britain's fishermen, boats and catch. There are 2939 fishing vessels over 10 metres in length in the British fleet: of these 1725 are in membership of the PO sector. While this represents only around 60 per cent of the total number of boats, in terms of gross registered tonnage (GRT) and total engine power (kW) the PO sector accounts for 80 per cent and 77 per cent respectively of the total UK fleet over 10 metres. In terms of fish quotas it is estimated that the PO sector manages over 95 per cent of all quotas. The 19 POs reflect the geographical and sectoral diversity of the British fleet. Details of the size and membership of each PO can be found in Figure 10.1 and Table 10.1.

That proportion of the British fleet which is not in membership of the PO sector is referred to as the 'non-sector'. Although the non-sector accounts for a fairly large number of boats, in quota terms it represents less than 5 per cent of British quotas. The non-sector is managed in much the same way as all fisheries were managed before 1984, i.e. by individual vessel monthly allocations set by fisheries departments. The non-sector largely consists of smaller vessels, but in addition to the non-sector, there is also a very large number (5372) of 'under 10 metre vessels' within the UK fleet. While very large in terms of numbers of individual vessels, this sector obviously consists entirely of small boats, many of which are operated on a part-time or seasonal basis, and is also managed directly by fisheries departments.

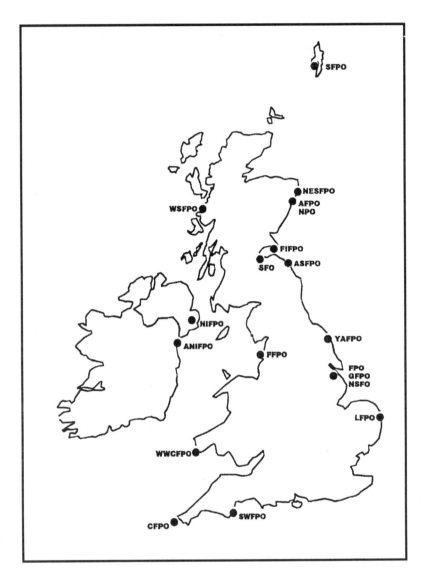

Figure 10.1   Producer organizations in the UK.

Table 10.1 Producer organizations within the UK in 1996

FPO	Year of recognition	No. of vessels over 10 m	Mean GRT	Total quota managed tonnes (1996)
Aberdeen FPO (AFPO)	1974	69	88	31 303
Anglo-North Irish FPO (ANIFPO)	1976	75	53	6 523
Anglo-Scottish FPO (ASFPO)	1975	138	37	19 392
Cornish FPO (CFPO)	1975	181	41	8 002
Fife FPO (FFPO)	1980	36	103	6 273
The Fish Producers Organization (FPO)	1973	43	379	10 162
Fleetwood FPO (FFPO)	1983	43	42	1 244
Grimsby FPO (GFPO)	1981	69	37	8 358
Lowestoft FPO (LFPO)	1993	11	286	4 272
North East of Scotland FPO (NESFPO)	1980	115	79	44 509
North Sea Fishermen's Organization (NSFO)	1993	40	226	11 185
Northern Producers' Organization (NPO)	1995	14	190	3 118
Northern Ireland FPO (NIFPO)	1976	160	51	20 943
Scottish Fishermen's Organizations (SFO)	1974	450	76	185 615
Shetland FPO (SFPO)	1982	65	241	59 566
South Western FPO (SWFPO)	1974	103	59	5 353
Wales and West Coast FPO (WWCFPO)	1993	47	227	5 696
West of Scotland FPO (WSFPO)	1995	51	20	2 125
Yorkshire and Anglia FPO (YAFPO)	1977	38	45	4 252

THE DEVELOPMENT OF SECTORAL QUOTA MANAGEMENT

During the past ten years, the SQM system has continued to develop and adapt to changing circumstances. Each year fisheries departments issue a consultation paper which proposes certain changes to the basic SQM system. On the basis of the response from the POs and the non-sector, changes are made to the SQM system for the ensuing year.

Through regular changes and modifications, the SQM system has now developed into a complex and comprehensive system of fisheries management. All POs must now manage all fisheries for which there are UK quotas in the North Sea (Area IV), West of Scotland

(Area VI) and the Irish Sea and English Channel (Area VII). It had previously been possible for POs to manage some fisheries and opt to remain under non-sector management (i.e. management by fisheries departments) for others. Sectoral quotas are currently calculated on the basis of the actual catches ('track record fishing performances') of member vessels during the previous three years. Pelagic quotas had previously been based on the catches of the previous two years. Since 1992, track record fishing performances have been attached to, and transferable with, vessel licences, rather than to the vessels themselves.

Having established the sectoral quotas available to each PO, fisheries departments then monitor uptake and will close a fishery when the sectoral quota has been caught. Apart from this the management of the quota is largely at the discretion of the PO concerned. As one would expect, different POs adopt different quota management strategies depending on their individual circumstances. In so far as white fish quotas are concerned, most POs continue to allocate quotas to member vessels on a monthly basis. Sometimes these quotas vary according to vessel size bands, but more often similar quotas are allocated to members regardless of vessel size. Several POs, however, have recently started to allocate individual annual vessel quotas based on the individual vessel licence track record fishing performance. In other words the quota allocation which a PO receives on behalf of a member vessel is simply reallocated to that vessel on an annual basis. Some POs allocate all their quotas on this basis (e.g. the Northern PO and the North Sea PO), others only for some species (e.g. the SWFPO) and others only for some vessels (e.g. the SFPO). Other POs (e.g. the FPO) have made similar arrangements but on an individual company basis as opposed to an individual vessel basis.

Pelagic quotas are essentially allocated to only two POs (the SFO and the SFPO) with individual pelagic quotas being allocated to those pelagic vessels not in membership of these two POs. Both the SFO and SFPO in turn allocate individual annual pelagic quotas to their pelagic members. Again these quotas are related to individual track record fishing performances.

A further development of the SQM system has been to allow POs complete discretion in swapping fish between each other. This ensures that UK quotas should not remain uncaught. Quota swaps are now becoming commonplace with POs swapping away fish quotas unlikely to be caught in return for fish quotas which are in short supply. During 1995, for example, there was a total of 250 quota swaps. Direct fish for fish quota swaps are most usual although quota gifts (which can be repaid in future years) are becoming more common. This quota swap system even allows for fish quotas to be effectively sold for cash (i.e. through a quota gift which is not re-paid in future years) but although this has happened, it is not usual. The purchase of quota from one PO by another PO will probably never become widespread because of the consequent permanent loss of track record by the PO selling quota. The quota swapping facility has, however, enabled the SQM system to become very flexible and responsive to the changing needs of industry.

Purchase of fish quotas has in effect been possible since 1993 through a special scheme which was introduced to complement the decommissioning scheme introduced at this time. Under this scheme, vessel owners interested in decommissioning their ships could, as an alternative, sell their quota entitlement (arising from their track record fishing performance) to the PO they had been in member-ship of for the past three years. In return for selling quota entitle-ment to the PO, the vessel owner had then to relinquish his vessel licence in the same way as when a vessel owner accepts a decommissioning grant. This scheme is essentially a form of industry-funded decommissioning, and since the requirement to have been in membership of a PO for three years has been relaxed, a vessel's quota entitlement can now be sold to any PO. The SFPO has probably used this scheme most, having purchased the quota entitlements of five separate vessels to date, with a view to owning these quota entitlements on a communal basis.

As already noted, fisheries departments continue to manage the non-sector. Over recent years the size of the non-sector has been reduced. Most larger vessels are now in membership of POs and the non-sector now consists of a large number of small vessels. The departure of larger vessels from the non-sector to the PO sec-tor has resulted in substantial track record fishing performance being lost from the non-sector. This has in turn resulted in the non-sector quota allocations being further reduced with very poor per-vessel quotas currently being allocated to non-sector vessels by Fisheries Departments.

As the SQM system has developed over the past few years, there has been a greater recognition of the importance of track record fishing performance. As each PO has endeavoured to maximize its sectoral quota allocation, increased attention has been focused on the catch record of vessels applying for membership. Most POs now have a policy of only admitting into membership those vessels which have a track record fishing performance comparable to vessels of a similar size already in membership. It is now widely recognized that admitting vessels with inadequate catch records will simply result in new members requiring quota allocations but unable to contribute significantly to the quota pool.

Within the last few years the non-sector has become a residual for that proportion of the fleet which has very poor track record catches and has consequently been unable to secure membership of a PO. Real fears have been expressed that, if more of the non-sector fleet with reasonable track record catches were to join the PO sector, the already poor non-sector quota allocations would become even worse. In response to this, fisheries departments have, since 1995, guaranteed a minimum quota allocation to the non-sector. This is based on the track record fishing performance of vessels in the non-sector for the three-year reference period from 1991 to 1993. Even if the actual track record of the non-sector falls below this historical level, fisheries departments have agreed to underpin allocations to the non-sector (and the under 10-metre sector) at this minimum level. This underpinning applies to all Area VII stocks together with some stocks in Areas IV and VI. If the non-sector or under 10-metre sector are unable to catch their 'underpinned' quota then Fisheries Departments reallocate this fish among the PO sector.

However, by essentially guaranteeing a minimum sectoral quota allocation to the non-sector, fisheries departments have actually introduced a fundamental change to the SQM system. No longer are sectoral quota allocations calculated strictly according to the historical catch records of member vessels. A sector of the fleet (the non-sector) is now guaranteed a minimum quota share (regardless of catch performance) of certain stocks. If this concept of fixed quota shares were to be extended to all sectors of the fleet, as is discussed later, the result would be a dramatic transformation of the SQM system.

SECTORAL QUOTA MANAGEMENT – AN ASSESSMENT

As with any system of fisheries management, the SQM system has advantages and disadvantages. Any assessment must analyse the advantages and benefits of the system against its disadvantages and costs.

There are a number of clear advantages of the SQM system. In European terms, it is a system which, to a large extent, ensures that the UK catches its quotas in full but does not exceed them. By and large, UK quota undershoots and overshoots are minimal. The SQM system is based on a total of 19 POs managing their individual quota allocations. A complex but effective system of penalties ensures that there are adequate deterrents to act as a disincentive to prevent the overfishing of individual sectoral quota allocations. As a result a significant UK quota overshoot is now improbable. In a similar way the quota swapping system ensures that there is sufficient flexibility to allow the UK, as a whole, to maximize uptake of all quotas available. As a result a significant UK quota undershoot is also now rather improbable.

A second advantage is that the SQM system operates on the basis of fisheries management by fishermen's organizations. This is important. Government allocates the annual sectoral quotas to the various POs (according to the established rules relating to track record fishing performance etc.) but it is then the responsibility of each PO to decide how the sectoral quota should be managed. Questions such as how the sectoral quota should be allocated between member vessels, what penalties should apply in the event of quota levels being exceeded and how the fishery should be managed over a 12-month period need to be resolved. These issues are generally dealt with by the board of directors of each PO, who are elected by the membership to this position. Whatever the outcome at the end of the year, the PO cannot blame anyone else for how fisheries are managed. Under the SQM system, responsibility for fisheries management lies firmly with the fishermen's organizations. This is one of the greatest strengths of the SQM system: the people who are most affected by fisheries management – the fishermen – have the principal responsibility for managing fisheries.

The devolution of fisheries management responsibilities to each PO also has the added advantage of allowing each organization to manage its own fisheries in a different way. Reference has already been made to the variety of quota systems adopted by different POs. In an industry as complex, geographically varied and different as fishing it is essential that appropriate quota systems can be adopted within diverse POs. The SQM system, unlike the quota management system which applied prior to SQM, enables this to happen. In other words the flexibility of the SQM system is another major advantage of this particular fisheries management system.

A further advantage is that the allocation of sectoral quotas affords POs the opportunity to more effectively market their members' catches. As already noted, the role of POs under the CFP is the improvement of the market for the catches of member vessels. The potential of approaching a buyer or several buyers with an annual quota of fish which can be landed at agreed prices is obvious. Notwithstanding this, few POs have actually realized the marketing potential which the SQM system provides. The exception has been the SFO which has, with some success, marketed its members' catches of prawns and pelagic species by exploiting the advantage of a total annual tonnage available for sale. The main reason why POs have not developed a corporate marketing approach would seem to be the attachment to the auction market which, particularly for white fish, continues to give member vessels a better return than any contract pricing arrangements.

The advantages of the SQM system are therefore clear. It is a flexible and dynamic system of fisheries management which has evolved over time. Above all, it is a system which has evolved with the full participation of the fishermen and is a system which is operated on behalf of fishermen by fishermen. This is its greatest strength. The very fact that the SQM system has survived during a period when most white fish quotas have been subject to a period of sustained reduction is probably the greatest testimony to this system's success.

The SQM system is not, however, without its disadvantages and problems. The single biggest criticism of the quota system in general, and SQM in particular, is the regular and widespread landing of over-quota fish or 'black fish'. It is widely recognized that there have been significant landings of over-quota fish in both the white fish and pelagic sectors for some time now. But this is not a consequence of any one system of fisheries management. The problem is that the total quota available is inadequate to sustain the existing fleet at current levels of profitability. There is a problem of excess capacity and this is a problem which the UK fleet shares with most other EU member states. Excess capacity relates not only to the physical size of the fleet but also to the fishing effort deployed by that fleet. However, whatever system of fisheries management was in operation, this underlying problem of excess catching capacity would still exist. The replacement of the SQM system by another system would not alter the fundamental problem of imbalance between resources and catching capacity. It is therefore quite erroneous to blame the 'black fish problem' on the SQM system or indeed any particular system of fisheries management. The 'black fish problem' is a reflection of fleet overcapacity and not the result of any particular system of fisheries management.

A more serious criticism of the current SQM system is the fact that quota entitlement is directly linked to historical catch performance during the preceding three years. In other words, the higher the recorded catch during the previous three years (the historical track record) the higher the resulting quota entitlement. For those vessels which are allocated an individual quotas from their PO directly related to individual quota entitlement, there is an additional clear incentive to maximize catches in order to maximize future quota entitlement. This results in a 'race for fish' and has even led to uncaught fish being recorded as landed (this is the opposite of 'black fish' and has been described as 'ghost fishing'). The growing recognition of track record fishing performance as the criterion for future potential quota entitlements is bound to escalate this problem.

This problem can also be seen in the value which is now placed on licences with large track record fishing performances. These licences are escalating in price and are now often worth more than the price of the boat itself. This is a trend which many fishermen deplore. The fact that some flagship operators from Spain and Holland are now openly buying licences with large track record fishing performances is highlighting this particular problem.

SECTORAL QUOTA MANAGEMENT – THE FUTURE

The SQM system has operated now for more than a decade. It has continually evolved and developed to meet new circumstances. However, it is now at a crossroads. Its future development could result in, to all intents and purposes, a system of individual transferable quotas (ITQs). Alternatively, a more radical development could result in communally owned quotas.

As discussed above, only a few POs allocate quotas to member vessels on the basis of individual track record fishing performance. In such circumstances there is clearly an incentive for these vessels to buy additional licences with good catch records. By aggregating these licences to their existing licence, vessel owners automatically improve the quota entitlement by the amount of catch record purchased. This is now happening with increasing frequency. If more, or most, POs should decide to allocate individual quotas based on individual catch records then a clear incentive will exist for vessels to improve their quota entitlement by purchasing track record fishing performance. Fish quotas will have then become an asset which can be bought and sold by the purchase or sale of a vessel licence.

There is currently much debate regarding the 'fixing' of track records. By fixing individual track records, the practice of ghost fishing could be eliminated since the incentive to record fish which have not actually been caught will have been removed. Individual vessel track records will have been fixed and will not vary regardless of what fish a vessel actually records as having landed. If track records were to be fixed in this way, it has been argued that vessels will become much more aware of their individual quota entitlements. This could result in a demand that this individual quota entitlement be reflected into an individual vessel quota. The consequence of this would be a system very close to ITQs.

Fisheries economists may argue that the system described above is not a system of ITQs, in the sense that fish quotas are bought and sold as in New Zealand. In New Zealand, fish quotas can be bought and sold as an individual asset. In the system described above, quota entitlement would still be purchased together with a vessel licence. Moreover, quota allocations are still made under the auspices, and with the consent, of each PO. Notwithstanding this, it is clear that the SQM system could develop into a system very close to ITQs if individual fixed quota shares were to be introduced and if more POs were to decide to allocate individual vessel quotas.

This is probably the most likely outcome if present trends are continued. The PO sector is already coming under more and more pressure to allocate individual quotas. Unless there are fundamental changes to the SQM system, it can be argued that the existing system is likely to develop into a system very close to ITQs.

The fishing industry as a whole has for long been hostile to the principle of ITQs as a fisheries management system. There has consequently been some considerable interest in the concept of fixed FPO sectoral quotas which has been suggested as an alternative to ITQs by the CFP Review Group. Under this proposal, individual track records would no longer exist; instead, each PO would obtain a fixed percentage share of UK quotas, and the quota would belong to each PO and not to individual vessels. This would represent a radical departure from the current system which is based on the concept of individual vessel track records.

The allocation of fixed quotas to POs would, as with fixed individual track records, remove the incentive to record fish which had not been caught. Apart from this, however, there are no other similarities with the fixed track system. By allocating quotas directly to POs, individual track records would be irrelevant, and the concept of quota entitlement, an individual asset which can be bought and sold, would no longer apply.

The fixing of PO quotas would effectively stop the emergence of an ITQ system and would instead establish a system of communally owned quotas. Such a system would confer very considerable powers on the PO sector. But steps would have to be taken to ensure that SQM remains flexible. For example, there would have to be an agreed quota transfer formula to allow individual vessels to move from one PO to another, since a change of PO is often necessary if a vessel is sold from one part of the country to another.

A shift from individual quota entitlements to communally owned fish quotas is not without precedent. The Shetland Ring Fence arrangement is essentially a system which ensures that a fish quota which has been purchased by the SFPO remains the property of the local PO and is not transferred to the track record of individual member vessels. Within the SFPO the ring fenced quota (i.e. that proportion of the SFPO total quotas which is owned communally) now accounts for some 20 per cent of that organization's cod quota. Another precedent is the underpinning of the non-sector quotas which has already been referred to; in a quite different way to the ring fence arrangement, the principle of underpinning also effectively results in a fixed sectoral quota.

The fixing of PO quotas would introduce for all POs the security currently provided by the ring fence and underpinning arrangements. Many within the fishing industry argue that a system of communally owned quotas would be more in keeping with the traditions of the British fishing industry than would a system of ITQs.

CONCLUSION

The SQM system has served the fishing industry well during the past decade. All fisheries management systems must, however, continually evolve in order to serve the changing needs of a dynamic industry. The SQM system is probably now at a crossroads. The system can either develop towards individual quota rights (ITQs) or else towards communally held quotas (fixed PO quotas). The debate as to which of the two systems should be adopted is now well underway within the fishing industry. Government ministers have indicated that they are prepared to respond to the wishes of the industry on this matter. Whatever the outcome of this debate, the decision on how the SQM system develops in the future will therefore be made by fishermen. This in itself reflects the greatest single strength of the SQM system – it is a system of fisheries management where decisions are taken by the fishermen themselves.

BIBLIOGRAPHY

CFPRG (1996) *A Review of the Common Fisheries Policy Prepared for UK Fisheries Ministers by the CFP Review Group*, 2 vols (London: MAFF).

Goodlad, J. (1993) 'Sea Fisheries Management: The Shetland Position', *Marine Policy*, 17, 5: 350–1.

Hatcher, A. (1996) *Devolved Management of Fish Quotas in the United Kingdom: Producers Organizations and Individual Quota systems*, paper presented to the Conference of the International Institute of Fisheries Economics and Trade in Marrakesh, Morocco, July.

Young, J., Smith, A. and Muir, J. (1996) 'Representing the Individual Fishermen: An Attitudinal Perspective on One PO's Membership', *Marine Policy*, 20, 2: 157–69.

11 Establishing Sea Territories: a Way Forward for the Development of the Fisheries of the European Union?

Kevin Crean

INTRODUCTION

This chapter examines the role of territoriality as a mechanism for establishing a stable fisheries management regime. The chapter draws on case study examples from the UK, European Union and Oceania that show how fisheries conflicts might be defused, how the relationship between the regulators and users might be improved, and briefly comments on the institutional changes that would be necessary to support this change of policy direction.

Over the last twenty years, it is difficult to recall an occasion when the fisheries of the European Union (EU) have captured the headlines conveying a message other than crisis or conflict and, more often than not, both. The major fish stocks upon which the EU depends have slipped beyond the brink of serious over-fishing (Symes, 1992) to the point where they are no longer sustainable. Despite intense regulation, unofficial sources suggest that maybe the equivalent again of the tonnage of fish caught in the 'common pond' of the EU is thrown away at sea (Crean and Symes, 1994a) and that a substantial proportion of illegally caught fish finds its way into the supply network (Commission, 1991). Concomitantly, the fishermen, the food industries that use fish and the maritime communities dependent upon fishing as a source of income and employment have all suffered. This is manifested in a sharp decline in the numbers of full-time fishermen, shortages of certain prime fish species and the depression or even disappearance of fishing as an occupation in some of our coastal communities (Crean and Symes, 1996).

Despite the relatively small size of fishing as an industry in the EU (Salz, 1991), especially in relation to the agriculture sector, the European Commission has made efforts to address the crisis and a variety of regulatory schemes have been tried in an effort to rescue the fish stocks. Unfortunately, despite often determined, and sometimes original, approaches to the problems of over-exploitation it has not proved possible to stem the tide of decline. Indeed, more often than not, the implementation of regulatory measures has exacerbated the problems: the stock crisis continues and the associated social fall-out has had adverse effects upon the resource users and the communities dependent upon fishing for their livelihood. An important casualty of the failure of the attempts by government to positively intervene in the crisis has been the credibility of the regulatory bodies in the eyes of the resource users. Thus as the crisis has deepened so has the rift between the regulators and the fishermen (Crean, 1994).

Clearly there are problems in the structure of the EU's fisheries that are deep-rooted and not easily remedied. The CFP itself, established on the principles of the Treaty of Rome, has been much criticized and singled out as the main source of difficulties. Indeed, to date, the CFP has failed to achieve its objectives: there has been little reduction in fishing effort, no evidence of stabilization – let alone recovery – of major food fish stocks and, as a result, no improvement in the economic returns to the fishing industry. The fundamental assumptions and basis of the CFP are in question, and there are many voices within the EU fishing industries calling for the policy to be radically modified or made redundant.

In the world at large the principles upon which the modern systems of fisheries management are based are being re-evaluated. Is it really possible to manage the fisheries sector using strategies based upon bio-economic theory? There are those commentators who evince that bio-economic theory fails to model in an appropriate way the true, unpredictable behaviour of fisheries in an uncontrollable, if not chaotic, ecological system (Holm, 1995). Instead, the models substitute a false notion of predictability for stock behaviour (and economic returns) upon which the operational principles of the CFP, total allowable catch (TAC) and quotas are based. The problems are compounded by the underlying political principles of the EU –

non-discrimination between member states, political neutrality and the application of policy measures which inevitably tend towards stalemate, notably the principles of 'equal access' and 'relative stability' (Wise, 1984). The Treaty of Rome calls for non-discrimination between member states, but how can fish stocks be managed in an 'equal access' regime that is clearly not working? The deployment of regulations (and yet more regulations) shows no sign of bringing the industries back from the edge of crisis. Furthermore the fishermen seem to feel justified in ignoring the regulations pointing out that their livelihoods are threatened and, contentiously, that they were not adequately consulted before the laws were introduced. All in all it appears that the CFP is a policy designed to maintain the status quo in a sector beset by intractable problems and seemingly forever threatened by adverse economic and environmental conditions.

So where do we go from here? There is a growing recognition, at least among social scientists, that we must look beyond bio-economic theory to lay the foundations of a workable management policy that can ultimately deliver a sustainable system of resource utilization (Crean and Symes, 1994b). They argue that the support of the fishermen (and their organizations) is essential if we are to make progress in management. Put another way, the social scientists believe that any policy dedicated to the 'solution' of fisheries problems will fail if it does not embrace the aspirations of the fishermen and their dependants.

Evidence from other fisheries exploitation systems in the world (Fong, 1994; Miller, 1989; Ruddle, 1989) has shown that where legitimate and recognized fishermen's institutions exist, then their involvement in the policy- and decision-making processes can lead to the adoption of more successful resource exploitation strategies. A number of the successful examples have at their core organizational structures and institutions in which management is devolved from government to the resource user groups. In return for the greater autonomy provided by government the fishing communities have, in most instances, demonstrated a more responsible attitude towards resource utilization and, in some cases, self-regulation. It also appears that where the devolved management feature is coupled with the existence of legitimate fishermen's organizations then there is the capacity to internalize the resolution of conflicts (Acheson, 1989; Hviding and Baines, 1992; Ruddle, 1987).

In specific terms, there are some particularly interesting examples from the fishing cultures of the island states of the Pacific, coastal communities have for centuries presided over ᴡᴇ̴ɪ̵-ᴏ̵ɪ̵ɢ̴ᴀ̴ɴized fisheries exploitation systems that appear to have achieved, at least until recent times, the elusive goal of sustainability (Johannes, 1978). It is believed that sustainability has been due, in no small measure, to the existence of a resource exploitation policy that has conferred, through the allocation of sea territory, management responsibility to the local fishermen. It would seem that 'ownership' of the rights of access to fisheries resources in a given sea territory has prompted a positive and responsible expression of collective action, free of many of the problems that currently afflict the fisheries of the EU.

While the territorial use rights in fisheries (TURFs) of the Pacific Basin are ancient and reflect cultures and customary legal systems quite distinct from those associated with European institutions, the principles governing the structure and modes of operation are potentially of great interest with respect to the fisheries problems of the EU. This chapter explores the lessons that might be learned from a study of territoriality with respect to the performance of the modern-day resource management function.

THEORETICAL ASPECTS OF SEA TERRITORIES

Dundas (1991) gives a formal definition of the term sea territory: 'a maritime area that is under the control of a nation state, and is subject through the jurisdiction of the state to the regulation of natural resource exploitation, environmental protection, monitoring, control and surveillance.'

A nation declaring a sea territory will seek to gain recognition, and validation in international law, for the boundaries and rights expressed by the territory. Usually the delimitation of the sea territory is dependent upon recognition within the law, thus the 0–6 and the 6–12 nautical mile limits off the coast of the UK were set up as a result of the Final Act of the European Fisheries Convention (1964). The control will extend to the characteristics of the exploitation systems in relation to the natural resources embraced by the territory, and in particular there will be the property rights function upon which the system of access is based. A further dimension to the control feature will be the focus on maintaining the integrity of the boundaries that define territory.

The definition of sea territory given above readily describes most of the examples of delimited maritime sea territories to be found in the 'Common Pond' of the EU, for example the sea territories delimited by the 0–6, 6–12 and 0–200 mile boundaries. The terms might also embrace the incidents of territoriality that arise as a result of the extra-legal 'appropriation' of favourable fishing areas in what may, in theory at least, be a common access system. The sea territory definition needs extension in the context when the rights of use and exclusion defined over a given territory are held collectively, for example by a coastal community. In this context, it is more appropriate to invoke the use of the term TURFs. Panayotou (1982) defines TURFs as

> community held rights of use (or tenure) and exclusion over the fishery resources within a specific area and for a period of time. Accompanying these rights might be certain responsibilities for maintenance and proper management of the resource base, as well as restrictions on the exercise of the rights of use and exclusion.

This extension of the character of sea territories is seldom encountered in the present-day fisheries of the EU but is not uncommon in other parts of the world (Doulman, 1993; Durrenberger and Pálsson, 1987; Kalland, 1996). Not least this is because within the concept of TURFs there is a very powerful local institutional element that welds the sea territory feature to a number of other characteristics (Pollnac, 1984). Thus the existence of the TURF is not only confirmed by local institutional control over a territory enclosed by a boundary but also by a 'policy' that may dictate: the access conditions to the TURF resources; the degree of transferability and divisibility of the rights associated with the TURF; the capacity of the TURF to contract or expand; and the use restrictions in terms of the specific resources and technologies directed at harvesting those resources.

The boundaries of a TURF manifest themselves in a variety of forms ranging from the physical to the notional yet in most cases confine a particular geographical area and the resources contained therein. In some instances the boundaries might relate to the operational location for a particular fishing method or gear, or the temporal or geographical distribution of a certain target species. The properties of the boundaries of the TURF might be likened to those of a biological membrane. Biological membranes around living cells display a semi-porous structure that permits selective diffusion across the membranes, and thus the cells may expand and contract in response to pressure differentials in the external and internal environments. Analogous processes occur with respect to TURFs: definition and diffuseness; permeability; expandability and divisibility (Pollnac, 1984; Christy, 1982). Thus TURFs have been found to change in geographical extent over time (Kalland, 1996). In some instances the right holders allow the entry of other users to their territory to prosecute resource utilization. This phenomenon brings in issues connected with the transferability of the right holding.

Overall the evolution of the properties of TURFs has in most of the examples that have come from the Pacific Ocean been presided over by organizations that have exhibited traditional and consistent institutional practices. Inevitably the traditional TURFs systems of the Pacific have reacted to contact with exogenous cultures and development processes. In some locations this has brought about the destruction of the system (Johannes, 1978; Doulman, 1993) while in others the TURFs are under threat as a result of contact with market-driven exploitation strategies and 'western' resource management regimes.

A BRIEF HISTORY OF THE DEVELOPMENT OF SEA TERRITORIES IN THE PACIFIC OCEAN

The archipelagos of the South Pacific Ocean began to support human populations at least four thousand years ago (Anon, 1990). The innate capacity of the islanders to travel great distances in quite small vessels led to a rapid colonization of all the key land masses and indeed some where life was, and to this day is, rather precarious.

The successful invasion of many of the island environments was followed by substantial increase of population and with it a growing pressure on the food resources of the island strongholds. There was no doubt an appreciation on behalf of the island communities that while the food resources of land and sea were dependable they were not unlimited (Johannes, 1978). Thus, in all probability, as a response to population pressure coupled with the need to husband scarce resources, the island cultures of the South Pacific developed systems for planning, developing and managing the coastal marine resources. This involved the division and apportionment of land and sea territories sometimes to individuals but more often to communities, together with the resources they contained (Christy, 1982; Cordell, 1984; Ruddle, 1989). The allocation of the TURF was made binding in customary law and was jealously guarded within the hierarchical social structures characteristic of the island societies (Sudo, 1984).

The boundaries of the sea territories were well known to the owners and their neighbours in contiguous areas, often demarcated by reference to topographical features such as islands, mounds of coral or sometimes wooden markers fixed in the reef. The boundaries were particularly relevant for the demarcation of areas where sedentary reef plants and animals such as seaweeds and molluscs were found. However, where more mobile, and even migratory, coastal marine species were concerned the boundary regulations could be eased to allow a more flexible harvesting strategy. Thus the systems were able to accommodate the harvesting of migratory species.

Conservation strategies were commonly used and embodied those technical measures that are familiar elsewhere in the world, including mesh regulations, prohibited gears and practices, closed areas and seasons, prohibited species, minimum landing sizes, protection of spawning individuals and the establishment of refugia.

The TURFs were not closed with respect to experimentation with, and acceptance of, new technology. Although for a long period of time the adoption of any new technology would have been in the context of a 'needs based' production system. Thus manufactured materials gradually replaced the traditional twines used in the making of fishing nets, and steel/plastic fishing lures have gradually been substituted for those made from mother of pearl and coconut fibres. In Ontong Java in the Solomon Islands an exotic form of post-harvest technology was introduced that enabled the islanders to prepare bêche-de-mer from sea cucumbers (Crean, 1977).

The institutional structure of the TURF was focused on the coastal community, which by definition included those who lived adjacent to the sea territory. The community members were subject to regulation through the customary laws and social code of the society. The community would display homogeneity in terms not only of race but also kinship, religion and often fishing practices. The sense of belonging, by whatever yardstick measured, would be strong (Pomeroy and Williams, 1994). In terms of the recent development of fisheries resources of the Pacific Island states there was only a tenuous link between the TURF managers and the national/regional fisheries authorities. In some instances the existence of TURFs was ignored by the fisheries authorities, as in the early days of commercial development of the Pacific fisheries attention turned to the largely offshore tuna resources. In others the system was only barely tolerated and sometimes viewed as an unnecessary constraint to more desirable commercial development (DPI, 1983).

Conflicts within (and between) TURFs were not uncommon, yet were for the most part resolved by reference to the law. The legal code was based in traditional law and was applied through the social hierarchy that dominated all aspects of life in the coastal communities (Ruddle, 1989; Johannes, 1978). Offenders who had contravened the laws of the TURF were subject to sanctions which might involve the payment of some form of compensation to the injured party but the use of imprisonment and physical punishment were not unknown.

For approximately two millennia the traditional marine systems of the South Pacific have presided over the orderly and apparently sustainable exploitation of marine resources; however, more recent contact with exogenous cultures has initiated a period of intense change. In some cases, the systems have experienced decline; indeed some have disappeared, or have been neutered by a combination of the influence of exogenous regulatory systems and contact with adverse economic and environmental forces.

Unlike other Pacific Island nations Japan has had a rather different history with respect to the development of territoriality in its inshore fisheries. Fishing territories were firmly defined in the Tokugawa period (1603–1868) in Japan and as the coastal waters were regarded as an extension of the land they too were part of the feudal domain. Control of the fishing territory was achieved through its management as a village estate under the control of the local community via the traditional hierarchy and powerbase. As in other Pacific Island states the boundaries were defined by recognizable topographic features.

The sea territories of Japan have continued to evolve even as we approach the end of the twentieth century. There are in particular some key developments that have shaped territoriality in a modern context. In the first instance a set of rights pertaining to territory owners was defined in 1901 and this statute was strengthened by the legitimizing of the right holder status by fishermen only as a result of the Fishery Law of 1949. This meant that individuals and private enterprises were able to obtain rights to fish only as long as the cooperative or fishermen group did not want to make use of their rights.

Furthermore, in 1948 (Ruddle, 1987) the fishing territories were redefined and the responsibility for their management given over to recognized fisheries cooperative associations (FCAs).

Over the centuries in coastal Japan there has been a tendency to develop larger territories. This appears to be a response to the inefficiencies of the smaller territories, particularly with respect to the economics of the marine products industries.

Overall the majority of the exploitation units in the coastal fishing territories are owner-operated and small-scale. A few licences have been issued to the more capital-intensive operators for fishing in coastal areas. The government authorities have taken action aimed at securing the livelihood of small-scale fishermen by placing restrictions on outside investments, thus preventing capitalized firms or individuals from usurping the access rights, or means of production, of the indigenous fishermen. There is a tendency therefore for the more heavily capitalized and larger-scale fishing operations to move out of the coastal territories into the offshore and distant water fisheries.

In summary then, TURFs in Japanese fisheries have existed for over one thousand years gaining legitimacy and definition in the feudal period, but withstanding the impact of a variety of influences to take their place as a cornerstone of fisheries planning, management and development policy in modern-day Japan (Ruddle, 1987). This has created a relatively stable climate for sustained production of marine biological resources. Kalland (1996) quotes the figure of between 2.5 and 3.0 million tonnes as being maintained annually over the last sixty years.

TERRITORIALITY IN THE UK AND EUROPEAN FISHERIES

Territoriality is a widespread and in some instances enduring condition of coastal fisheries exploitation and management in world fisheries. While the phenomenon is not unknown in UK and European fisheries its expression has been limited largely by the influence of the prevailing institutional structures, as well as the national and European Union legal systems.

In the UK there are formal expressions of territoriality made by the conditions of access that prevail in the 12-mile zone. The European Fisheries Convention of 1964 created the six- and 12-mile seaward limits and these boundaries were subsequently absorbed within the territorial limitations of the CFP. The control of resource exploitation within the six-mile limits is vested in the UK authorities with regulations created and applied by a combination of the sea fisheries committees (in England and Wales) and the MAFF. There is no formal ascribing of management or exploitation 'rights' to particular fishing groups, although it is sufficiently close inshore for informal (and may be extra-legal) conditions to operate. In theory the 0–6-mile limits, currently existing as a derogation of the CFP, might be withdrawn in the year 2002.

Similarly the 6–12-mile limit was established in the mid-1970s and has undergone some changes since its incorporation within the CFP. The resources within the 6–12-mile limit are accessible to certain groups of European fishermen who together with their British counterparts have traditionally fished in these waters. The boundary is not always 'permeable', and in areas off eastern England and north and west Scotland for example fishing activity by non-UK fishermen is not permitted. The 6–12-mile territory, because of its greater distance from the coast and access by a more varied group of users, presents greater problems of management and control.

Beyond the 12-mile limit there is a system of ground closures and exclusivity devices which in the name of resource conservation exert quite distinct territorial effects on the user groups active in the 'common pond'. Seasonal ground closures have been instituted to protect spawning or nursery stocks as in the case of the herring prohibitions on the North and Irish Seas. Likewise the much larger Mackerel Box off the south-west coast of England seeks to protect juveniles of the western mackerel stock by the somewhat indirect means of restricting the amount of mackerel on board to 15 per cent of the total catch except in the case of hard lining. In general these measures, although creating differentials of benefits between user groups, are accepted as being in the interests of stock conservation. However, the sea territory defined by the Norway Pout Box in the northern North Sea has proved more contentious. The aim of the box was ostensibly to reduce the amount of industrial fishing in the sea territory where juvenile stocks of haddock and whiting are abundant, but in practice the establishment of this zone has discriminated against Danish industrial fishermen.

The accession of the UK to the CFP was accompanied by the gradual development of certain restrictions and controls on access that have given rise to stronger expressions of sea territory. One of the most interesting and potentially instructive examples is that of the North of Scotland Box. This territory embraces the coastal waters of northern Scotland and the Orkney and Shetland Islands. The Box represents the only concession granted to the UK in response to its demand for preferential access beyond the 12-mile limit at the time of the renegotiation of the CFP, 1976–82. In principle the main criterion for restricting access within the Box was to grant preference to local fishing vessels; in practice, however, the focus on the limitation of vessels over 26 m has actually allowed access to large numbers of UK vessels fishing out of non-local ports. The Box is, in fact, a Community fishing zone with a specific licence allocation for vessels over 26 m and is managed by the Commission rather than the UK authorities. In effect a coastal sea territory has been established which, if it survives 2002, could serve as a focus for greater definition of its boundaries and a changing permeability to the groups of users that currently exploit the resource therein. In many ways the users and water embraced by the North of Scotland Box represent one of the most obvious examples of a fisheries-dependent region – a concept that has been discussed and in some limited ways defined during recent developments of the CFP. The extension of oil and gas prospecting and the movement offshore of aquaculture operations could significantly change the character of fish production activities within the Box. It is interesting to note that the Shetland Islands Council is examining the feasibility of extending its jurisdiction over shellfish resources out to the 12-mile limit around the Shetlands. It is apparent that the box does offer some interesting prospects for development in terms of the role of user groups within the boundaries, although in the short term it is unlikely that there would be a further shift of emphasis favouring local participation in fish capture activities, to the exclusion of groups of users from outside of the territory.

More recent research within the UK (Symes et al., 1995) has raised the possibility of the management of fisheries resources embraced within sea territories defined by the existing boundaries of the ICES fisheries management areas. This research has concentrated on investigating the revised institutional structures and arrangements that would be necessary to manage sea territories of this type. The research concentrated mainly on evaluating the capacity of the UK institutions to manage some of the ICES areas in the NE Atlantic.

Clearly all the preceding examples of territoriality that are derived from UK (European) fisheries contrast vividly with the sea territory concepts that have arisen from the Pacific Island nations. If we consider how the development of sea territory in UK fisheries might be guided to improve the performance of the coastal fisheries exploitation systems then a number of problems arise which would need to be overcome.

Could the concept of fisheries-dependent regions be a starting point for the delimitation of sea territories in which the contiguous coastal communities acquire the benefits accruing through preferential access? While there is a theoretical appreciation of the importance of supporting regions that are dependent upon fishing, in practice it has proved difficult to define 'fisheries dependency'. Some argue that where fishermen exist then you have fisheries dependency. Furthermore any discrimination between groups of users within regions would have to be framed in the context of the prevailing common market conditions as well as the powerful features of the CFP that tend to work against regionalism in fisheries – equal access, relative stability, etc. However, it might be possible to elicit further support for the fisheries-dependent regions if we invoked the concept of subsidiarity which in a fisheries context would favour a regional devolution of management. Nevertheless to make progress in the empowering of fisheries-dependent regions through the process of subsidiarity would require the identification and strengthening of 'regional' organizations that could effectively manage fisheries within the overall demands of the CFP. An immediate difficulty in the UK in identifying organizations capable of undertaking regional management arises with respect to the number and diversity of organizations involved in fish production in any one location. There is a marked dissimilarity in orders of magnitude of organizations in their commitment to the principles of fisheries management. Furthermore it is difficult to identify existing organizations that currently exhibit a regional coherence in terms of the spread and 'boundaries' of membership. This contrasts vividly with the organizations that are characteristic of the fisheries exploitation systems of the Pacific.

The UK producer organizations (POs) have displayed the capacity to manage fisheries as evinced by their track record in sectoral quota management. However, this is to some extent negated by trends towards increasingly unstable membership and a tendency for existing POs to fragment. In general fishermen's federations and vessel owners' associations, while often having a geographical consistency in their membership, lack any organizational experience in the fisheries management process. In many ways the Sea Fisheries Committees, or a management structure derived from their present configuration, might well be a better starting point for the development of a management unit to administrate sea territories. On the plus side they have a geographical rationale with clearly defined seaward and lateral boundaries. In addition they are organizations that have experience of the management of shellfish resources and the legal instruments necessary to achieve this. On the negative side they are regarded as being rather weakly constituted, with a restricted local user participation. The SFCs do not exert management control over fin fish stocks nor do they exist beyond England and Wales in the UK.

CONCLUSION

There are numerous examples of 'successful' coastal fisheries management systems among the Pacific Island states although in some locations the pressures of commercial resource exploitation have contributed to the decline of the effectiveness of community-based management. It appears that where strong local organizations exist and are supported in law by the national government then the distorting effects of market forces can be effectively resisted. Given the parlous state of the fisheries of the EU it would be instructive for regulatory authorities to look more closely at how the national organization of fisheries exploitation might be improved. A potential solution might be through the delimitation of sea territories where regional user groups take a more active role in the planning, development and management of fisheries in return for preferential access.

BIBLIOGRAPHY

Acheson, J. M. (1989) 'Where Have All the Exploiters Gone?', in Berkes, F. (ed.), *Common Property Resources – Ecology and Community-Based Sustainable Development* (London: Belhaven Press), pp. 199–217.

Anon. (1990) *Third World Guide 1991/92: Montevideo* (Montevideo: Instituto del Tercer Mundo).

Christy, F. T. (1982) *Territorial Use Rights in Marine Fisheries: Definition and Conditions*, FAO Fisheries Technical Paper 273 (Rome: FAO).

Commission of the European Communities (1991) *Report 1991 from the Commission to the Council and the European Parliament on the Common Fisheries Policy* (SEC (91) 2288 final, Brussels).

Cordell, J. C. (1984) 'Defending Customary Inshore Sea Rights', *Senri Ethnological Studies*, 17: 301–26.

Crean, K. (1977) 'The Bêche-de-Mer Industry, Ontong Java, Solomon Islands. South Pacific Commission', *Fisheries Newsletter*, 15: 36–48.

Crean, K. (1994) *Social Objectives and the Common Fisheries Policy*, paper presented at workshop entitled 'An Agenda for Social Science Research in Fisheries Management', The Borschette Centre, Brussels, 5–6 May.

Crean, K. and Symes, D. (1994a) 'The Discards Problem: Towards a European Solution?', *Marine Policy*, 18, 5: 422–34.

Crean, K. and Symes, D. (1994b) *Social Objectives, Social Research and the Recalibration of Management Policies in Fisheries: The Case of the European Union*, paper presented at the ICES 1994 Annual Scientific Conference, St John's, Newfoundland, 20–30 September 1994.

Crean, K. and Symes, D. (eds) (1996) *Fisheries Management in Crisis: A Social Science Perspective* (Oxford, Fishing News Books).

Department of Primary Industry (1983) *Coastal Fisheries Development Plan*. Government of Papua New Guinea, Port Moresby.

Doulman, D. J. (1993) 'Community-Based Fishery Management: Towards the Restoration of Traditional Practices in the South Pacific', *Marine Policy* 17, 2: 108–117.

Dundas, C. W. (1991) 'Maritime Boundary Delimitation Negotiations', paper presented at the Training Course in Fisheries Management and Development, Antigua and Barbuda, September 1991.

Durrenberger, E. P, and Pálsson, G. (1987) 'Ownership at Sea: Fishing Territories and Access to Sea Resources', *American Ethnologist* 14, 3: 508–22.

Final Act of the European Fisheries Commission (1964) *European Fisheries Convention*, (London: HMSO, Cmnd 2355).

Fong, G. M. (1994) 'Case Study of a Traditional Marine Management System: Sasa Village, Macuata Province, Fiji' (*Project RAS/92/T05 Case Studies on Traditional Marine Management Systems in the South Pacific*. South Pacific Forum Fisheries Agency, FAO, Rome).

Holm, P. (1995) 'Fisheries Management and the Domestication of Nature', paper presented to the Fifth Annual Common Property Conference 'Reinventing the Commons', Bodø, Norway, 24–28 May, 1995.

Hviding, E. and Baines, G. B. K. (1992) 'Fisheries Management in the Pacific: Tradition and the Challenges of Development in Marovo, Solo-

mon Islands', Discussion Paper 32, United Nations Research In\
for Social Development, Geneva.

Johannes, R. E. (1978) 'Traditional Marine Methods in Oceania and th\
Demise', *Annual Review of Ecology and Systematis* 9: 349–64.

Kalland, A. (1996) 'Marine Management in Coastal Japan', in Crean, K.\
and Symes, D. (eds), *Fisheries Management in Crisis: A Social Science Perspective* (Oxford: Fishing News Books), pp. 71–83.

Miller, D. L. (1989) 'The Evolution of Mexico's Spiny Lobster', in Berkes, F. (ed.), *Common Property Resources – Ecology and Community-Based Sustainable Development* (London: Belhaven Press), pp. 185–98.

Panayotou, T. (1982) 'Territorial Use Rights in Fisheries', paper presented at the FAO World Conference on Fisheries Management and Development, Rome: 6–9 December.

Pollnac, R. B. (1984) 'Investigating Territorial Use Rights among Fishermen', *Senri Ethnological Studies*, 17: 285–300.

Pomeroy, R. S. and Williams, M. J. (1994) *Fisheries Co-management and Small-scale Fisheries: A Policy Brief* (Manila: International Center for Living Aquatic Resources Management).

Ruddle, K. (1987) 'Administration and Conflict Management in Japanese Coastal Fisheries' (FAO Fisheries Technical Paper 273, Rome, FAO).

Ruddle, K. (1989) 'The Organization of Traditional Inshore Fishing Management Systems in the Pacific', in Neher P. A., Arnason, R. and Mollet, M. (eds), *Rights Based Fishing*, (Dordrecht: Kluwer Academic Publishers), pp. 73–85.

Salz, P. (1991) *The European Atlantic Fisheries: Structure, Economic Performance and Policy* (The Hague: Agricultural Economics Research Institute, Fisheries Division).

Sudo, K. (1984) 'Social Organization and Types of Sea Tenure in Micronesia', *Senri Ethnological Studies* 17: 203–30.

Symes, D. (1992) 'The Common Fisheries Policy and UK Quota Management', *Ocean and Coastal Management* 18: 319–38.

Symes, D. and Crean, K. (1992) 'Alternative Management Systems for Marine Fisheries: Prospects for the North Atlantic Region', in Suarez de Vivero, J. L. (ed.), *The Ocean Change: Management Patterns and the Environment* (Seville, University of Seville), pp. 242–52.

Symes D., Crean, K., Mohan, M. and Phillipson, J. (1995) 'Alternative Management Systems for the UK Fishing Industry', Working Paper 6, compiled for the European Commission: *A Comparative Summary Report on the Alternative Management Systems for Denmark, Spain and the UK. Devolved and Regional Management Systems for Fisheries.* EU Funded Research Project No. AIR-2CT93-1392: DG XIV SSMA.

Wise, M. (1984) *The Common Fisheries Policy of the European Community* (London: Methuen).

...ds 2002:
...iarity and the
...gionalization of the
Common Fisheries Policy
David Symes

INTRODUCTION

In the run up to the year 2002, the search for the formula that might assuage the swelling chorus of disapproval for the existing Common Fisheries Policy (CFP) is gathering pace. Close analysis of the criticism would suggest that the principal weakness of the present policy derives very largely from attempts to impose a single, monolithic structure upon what is a large, highly diversified and ecologically sensitive 'common pond' stretching through some 30° of latitude from the innermost reaches of the Gulf of Bothnia to the Straits of Gibraltar. This Leviathan approach is combined with a failure to take account of the impacts of structural and conservation policies on the socio-economic conditions in those coastal regions which depend most heavily upon fisheries for their livelihoods. Much of the complaint is directed not at regulatory instruments *per se*; indeed, there is a general consensus throughout the industry of the need to reduce harvesting capacity and to regulate fishing effort. Instead criticism is focused largely on the failings of the institutional framework – its lack of sensitivity, its failure to incorporate professional advice and experience from the resource users and the allegedly uneven implementation of policy among the different member states. The quest, therefore, is for a policy system which will guarantee greater regional sensitivity in its approach both to the management of a sustainable resource base and to the sustainable development of fisheries-dependent regions.

The far sighted 1991 Review (Commission, 1991) looked forward not so much to the imminent mid-term assessment of the CFP but rather to the period leading up to the construction of a new common fisheries policy after the present one lapses in 2002. The Review contained a number of hints as to how fisheries policy might be reconstructed around the interlinking of subsidiarity, co-management and regionalization. None of these ideas were elaborated within the Review, nor has there been substantive evidence of a shift in those directions in the period since its publication.

In looking to alternative ways of reconstructing fisheries policy after 2002, three contrasting approaches can already be identified. First, there are those who favour the status quo, believing that the current system offers the best possible compromise between recognition of the fundamental principles of the EU, as defined in the Treaties of Rome and Maastricht, and the challenge of managing the Union's common fisheries; they would seek only minor modifications to the existing system. There may even be some who would wish to see further centralization of decision-making in order to reduce the potentially contentious discrepancies that can arise from differences in implementation and level of enforcement on the part of individual member states.

Second, there are those who would dismantle the CFP, either through unilateral withdrawal and the repatriation of the resources within the 200-mile limits and reversion to management through the unfettered actions of the nation state, or through the prosecution of 'coastal state management' within a much reduced common framework.

Third, there are those who seek a fundamental reform of the CFP. While acknowledging the need to retain a set of common principles governing the exploitation of fisheries within the 'common pond', they seek the implementation of some form of devolved management by which to regain the commitment and willing compliance of the resource users. Among this group are those who advocate a regionally devolved approach to fisheries management. The regionalization of fisheries policy is becoming an increasingly fashionable theme. One of the few really positive recommendations for the reform of the CFP to emerge from the UK Ministerial Review Group was for the promotion of Regional Consultative Committees (CFP Review Group, 1996). But despite its growing popularity, there is as yet little substance to the concept of regionalization.

The aim of this chapter is fourfold: first, to relocate the notion of regionalization within the broader context of devolved management; second, to review the existing evidence of regionalization within the EU fisheries policy; third, to propound the concept of

management' as an appropriate option for the fu-
ent of fisheries management in the NE Atlantic; and
lore the possibility for a closer integration of policies
achieve sustainable fisheries with those intended to
economic and social development of fisheries dependent regions.

DEVOLVED MANAGEMENT

According to a Draft Opinion from the Committee of the Regions:

> It may be that the present management system which is essen-
> tially that of command and control from top down is not fully
> seen as legitimate from the point of view of the individual fishermen
> and particularly ... of regions which are highly dependent on
> fisheries.   (COR, 1995a)

Devolved management addresses the problem of legitimacy by
interweaving three strands of thought which, by and large, are missing
from the existing management regime for the EU's fisheries, namely
subsidiarity, user participation and regionalization. Subsidiarity is
already recognized as a basic principle for the further development
of EU policies; the idea is enshrined in the Treaty of Maastricht,
Article 3b, which states that:

> the Community shall take action, in accordance with the prin-
> ciple of subsidiarity, only if and in so far as the objectives of the
> proposed action cannot be sufficiently achieved by the member
> states and can, therefore, by reason of scale or effects of the
> proposed action, be better achieved by the Community.

Put more simply, subsidiarity, also referred to as the 'proximity
principle', implies that decisions should be taken at a level as close
as possible to the citizen. It embodies the notions of *democratic
legitimacy* by avoiding excessive centralization of power; *transparency*
by encouraging a clear allocation of functions between the different levels of decision-making; and *efficiency* since it presupposes
that powers are exercised at the most appropriate level (COR, 1995b).
To date, within fisheries management subsidiarity applies principally at the level of implementation where it is the member state
that must ensure that the rules are properly policed by national
inspectorates and enforced through national courts.

It follows that decisions which recognize these three qualities – democratic legitimacy, transparency and efficiency – must also rely upon admission into the decision-making process the experience and advice of those professionally engaged in the field of activity which is to be managed, implying the incorporation of responsible fishermen's organizations within the policy process. Much has been made of the concept of co-management (Jentoft, 1989; Jentoft and McCay, 1995) as providing an ideal framework for decision-making within fisheries. As yet, however, the involvement of user group organizations within policy-making in Europe is characterized by considerable differences in national political cultures, ranging from the so-called 'negotiation economies' of the Nordic countries (see Nielsen and Vedsmand, 1998) where the custom has been for professional organizations to be consulted formally at an early stage, to the centralized bureaucratic decision-making characteristic of the UK (see Phillipson, 1998) where professional organizations have been largely excluded from policy discussions until a very much later stage.

Regionalization does not automatically imply the inclusion of the subsidiarity principle nor an invitation to professional organizations to join in policy discussions. There is a need to distinguish between the situation where the central administration imposes decisions from above which discriminate between regions in a quantitative and qualitative way and the idea of regional self-management, informed by grass-roots opinion and reliant on endogenous powers of decision-making through local or regional democratic institutions. Devolved management, in its pure form, seeks to involve the delegation of powers to responsible organizations whose constitution reflects the proximity principle with respect to both regional representation and professional competence.

REGIONALIZATION OF FISHERIES MANAGEMENT

The concept of regional management is certainly not new to fisheries, nor has the CFP been totally immune from regional influence. For a number of years in the postwar period, there was a vain attempt to manage the North Atlantic fisheries through two ill-fated international commissions – the International Commission for Northwest Atlantic Fisheries (ICNAF) in the North West Atlantic and the North-East Atlantic Fisheries Commission (NEAFC) in the North East. Their failure can be explained by incomplete membership; an unwillingness on the part of member states to accept scientific advice and set appropriate precautionary quotas; and a lack of enforcement capability. The first and last of these problems have also frustrated the attempts by the North Atlantic Fisheries Organization (NAFO), as the successor to ICNAF, to manage the high seas fisheries in the NW Atlantic in more recent times.

As Wise (1996) points out regional demands have already played an influential role in shaping the design of the CFP. The struggle between the 'fish rich' members of the EC9 and the 'fish poor' members in the renegotiation of the policy in the late 1970s and early 1980s resulted in a number of regional concessions to the principle of open (or equal) access. Among the most important were: the derogation allowing recognition of national territorial waters out to 6 or 12 nautical miles for the benefit of those vessels 'which traditionally fish in those waters and which operate from ports in that geographical coastal area' (Article 100, 1972 Treaty of Accession); the Hague resolution, intended to acknowledge the special needs of fishing-dependent regions in Ireland, north Britain (and, originally, Greenland) through preferential application of quota allocations in certain circumstances; and access to the Shetland Box, which granted preferential access to local fishermen through a restricted licensing system, and the Irish Box – a temporary measure intended to moderate the initial impact of Spain's accession to the Community. But these discriminatory actions were to prove little more than anomalies in the general configuration of a CFP based on equal access and non-discrimination – exceptions to the rule, rather than the rule itself.

However, the clearest example of a regional dimension to the current policy, and one which concerns the cornerstone of the management regime, is the use of ICES fishing areas for the calculation and allocation of TACs and national quotas. In a very real sense, therefore, existing policy can be said to be constructed on regional lines – but only through a mechanism of centralized decision-making and the top-down delivery of policy decisions, with little or no direct consultation among relevant regional organizations.

There have been hints, too, of a regional approach to structural policy. The Multi-Annual Guidance Programme (MAGP) has evolved through three generations to a point where the latest Programme (1993–6) has set very much more specific sectoral targets than its predecessors, though without taking account of its socio-economic impacts in particular regions. The intention for the fourth MAGP (1997–2000) was that it should combine a sectoral and a regional approach, setting targets for different fishing areas. The Lassen Report (1996), leading up to the next round of proposals, provided a very detailed analysis of the state of the stocks and the distribution of effort by different types of fishing activity for each of the ICES subdivisions. The first indications of the new MAGP proposals (*Fishing News*, 1996) suggest that while objectives will be identified for each of these subdivisions, there will be little variation in the aggregate targets set. It also seems likely that the new proposals for technical conservation will likewise revert to broad, horizontal measures applying throughout much of the 'common pond', rather than reflect the kind of regional sensitivity captured by the NFFO's proposals of 1993 or in the UK government's discussion paper (MAFF, 1996).

FISHERIES-DEPENDENT REGIONS

As a framework for planning and development, the region is essentially a 'terrestrial' concept. The EU has developed a comprehensive and well integrated regional policy, which includes several instruments for assisting fisheries-related development projects. Community funding is available, through the Financial Instrument for Fisheries Guidance (FIFG), for the adaptation of structures in fisheries under Objective 5a of the integrated regional development policy. The sector is also eligible for financial assistance from the European Regional Development Fund (ERDF) and European Social Fund (ESF) in regions covered by Objectives 1, 2 and 5b; in effect, most of the Community's fisheries-dependent regions are included under one or other of these definitions. To date, however, the socio-economic impacts of fisheries policy have been treated as an external cost and thus dealt with through other policy areas, rather than as a primary consideration for the objectives of fisheries management itself (Symes and Crean, 1993).

More recently attempts have been made to define and characterize fisheries-dependent regions more precisely (Commission, 1992) so as to provide a geographical framework for the PESCA initiative, dedicated to helping the industry respond to the need for restructuring through the creation of alternative employment opportunities.

Emergency aid is required in those regions most seriously affected by some of the more draconian measures intended to achieve a sustainable resource base, including the rapid downsizing of the fishing fleets and sharp reductions in TACs for key target species. But at present the linkages between conservation and structural policies, on the one hand, and the spatial distribution of their socio-economic impacts on the other are not sufficiently well understood. Technological development within the industry and the increasing mobility of fishing fleets in search of diminishing resources have tended to sever the traditionally close links between fishing communities and local fishing grounds for all but the inshore sector.

Although the responsibility for regional projects is commonly vested in national, regional and local governments, possibly acting in partnership with private capital, there is a role for voluntary associations, drawn principally from grass-roots organizations, to act as catalysts for development. Only rarely will these assume a transnational character. One example from within the EU is the *Arc Atlantique*, a voluntary association of regional authorities along the Atlantic Coast (see Figure 12.1), which seeks to encourage inter-regional cooperation through a more efficient exploitation and management of indigenous resources in an area sharing common characteristics, interests and problems defined by its peripheral, maritime location (Wise, 1996). Thus far, few concrete results have emerged, but the association provides an opportunity for joint action not only for common development projects but also for compromise solutions to shared international fishing disputes, providing there is sufficient will among fishermen's organizations, which are unfortunately more usually in conflict over the defence of their local interests.

REGIONAL SEAS MANAGEMENT

Coastal state management is posited by some as the only viable alternative to the monolithic approach of the existing CFP. But such a claim is specious. In truth, neither approach is particularly appropriate, in scale or structure, to the complex geopolitical conditions of Europe's fisheries. The 'common pond', stretching from the Gulf of Bothnia to the Straits of Gibraltar, is clearly far too large and diverse to suit a single, uniform management regime (Figure 12.2). But equally, given those same conditions, coastal state management is unrealistic in the sense that most fisheries in the semi-

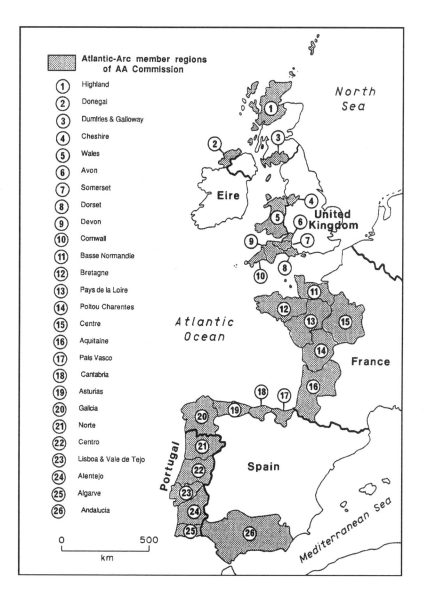

Figure 12.1 The 'Arc Atlantique' (based on Wise, 1996).

Baltic, North and Irish Seas are shared between several states (Figure 12.3) and therefore require some overarching c management to secure the sustainability of the stocks and uitable allocation of resources. The lessons of the ill-fated international commissions – albeit operating under significantly different circumstances in the days before exclusive fishing zones – suggest that binding and enforceable agreements are necessary. Effective management in such crowded and overexploited waters requires the kind of legal framework and discipline that only a formal, parastatal organization like the EU can provide. But this is certainly not to argue for the retention of the CFP in either its existing geographical shape or its current political structure.

Attempts to formulate an effective system of regional management face a particular dilemma caused by the indivisibility of marine ecosystems, on the one hand, and the need to establish a *modus operandi* which can channel the common objectives of different resource user groups into patterns of behaviour characteristic of 'good neighbours', on the other. The regional approach is driven not by any ecological imperatives but by the need for resource users to work together for the common good. The present system of centralized 'command and control' management over such a large 'impersonal' space does not allow this to happen. The 'regional seas' approach may offer a solution.

What is proposed[1] is a decentralization of management through the creation of a comprehensive series of Fisheries Councils covering the EU's 'regional seas'. These are easy enough to define in respect of the enclosed or semi-enclosed seas (Baltic, North, Irish and Mediterranean) but are less easily resolved in areas of 'open' seas like the West of Scotland, Channel, Bay of Biscay, etc. A ready-made solution is available through the use of ICES divisions to define the boundary lines (see Figure 12.4). The ICES framework is certainly not ideal: the arbitrary subdivision of ocean space into geometric designs makes no pretence at drawing boundaries around discrete ecosystems. It does, however, have the merit of providing a reasonable scale of operation, of eschewing prejudicial notions of territoriality and of conforming to frameworks used for biological advice and TAC allocations.

Fisheries Councils would be given appropriate responsibilities and powers for making recommendations concerning the detailed management policies within the relevant regional sea, via the Commission, to the Council of Ministers. All aspects of fisheries

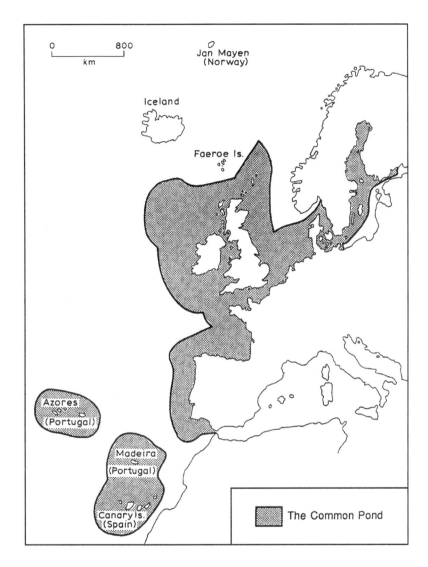

Figure 12.2   The 'common pond' (based on Wise, 1996).

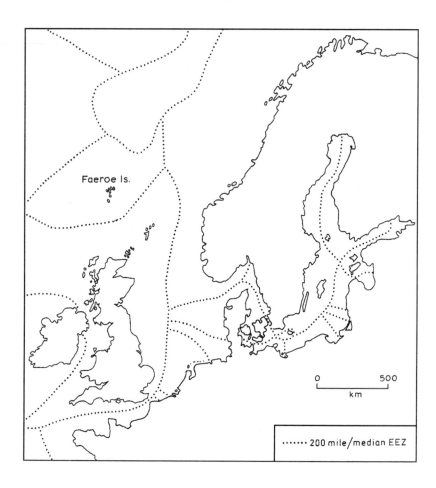

Figure 12.3 Northern Europe: coastal state EEZs.

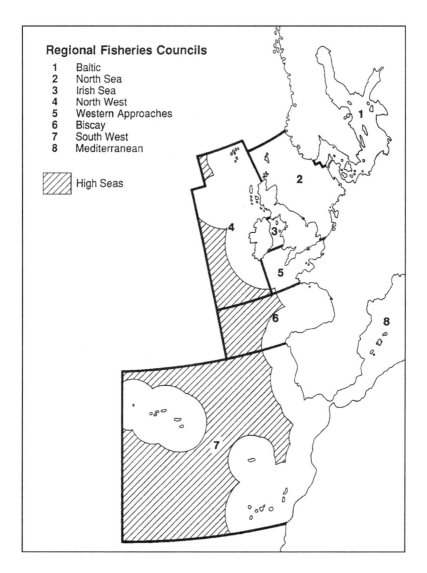

Figure 12.4 Regionalization of Europe's Common Fisheries Policy: Regional Fisheries Councils.

management – quota and/or effort allocation, gear regulations, ground closures, structural reform and market regulation – should be determined, in the first instance, by the Regional Councils. In formulating their proposals, the Councils should also have due regard for the socio-economic impacts in those areas with a particular dependence on fishing-related activities. The Councils might also have a role in coordinating strategic policies for the development of such fishing-dependent regions with a possible restructuring of the PESCA initiative along regional lines. Such policy proposals would need to be consonant with the broad aims and objectives of the CFP as a whole (and with the FAO's guidelines for responsible fishing). The regulatory measures would normally be developed from a broad package defined by the CFP, but there would be scope for variation where particular conditions of the region warranted.

The constitution of the Fisheries Councils needs to be defined both geographically and professionally. 'Political' membership would be reserved for the appropriate coastal states and other countries with established fishing rights, as currently identified by the allocation of TACs; observer status, without voting rights, could be granted to other interested parties outside the membership of the EU. In the quest for greater participation from resource users, delegations would comprise not only scientists and administrators but also representatives of the relevant fishermen's organizations and be led by an appropriate government minister. Membership of the councils could be formed of representatives from either the member states involved (central administrations, national fishermen's organizations) or the relevant regional administrations and professional organizations. The former seems the more likely solution as there is, as yet, no formal recognition for regional administrations to act in lieu of member states and, indeed, not all member states have appropriate regional authorities; but the latter would come much nearer to satisfying the proximity principle of bringing policy-making closer to the resource user.

Chairmanship of the councils could rotate among ministers from member states with a major stakeholding in the regional sea. While it would be preferable for any recommendations to the Council of Ministers to be reached on the basis of consensus, it is likely that, in reality, a system of qualified majority voting (QMV) would be needed on some, more contentious, issues. The system should reflect the relative weighting of the importance attached to the specific regional fishery within the national fishing industry as a whole.

Constructing a workable key to both membership and, more especially, voting rights within the councils is certain to prove problematic. Membership could be defined in two 'bands': band A would assume coastal state priority in the allocation of voting rights based on a combination of the total landings and the share of the national catch derived from region 'x', measured either in 'cod equivalents' or first-hand sales value; band B members – non-coastal states with established historic rights in region 'x' – would also be assessed on a similar basis, but at a significantly lower tariff. It may be necessary to create a third tier of representation to accommodate those countries, not qualified by coastal state or historic rights definitions, which have a common interest in the exploitation of a transboundary stock in a neighbouring region.

Further advantages can be ascribed to the 'regional seas' approach to fisheries management in Europe. In the first place, the system could be built up from (or, alternatively, disaggregated into) national arrangements for the regional administration of fisheries. In the UK, for example, the CFP Review Group (CFPRG, 1996) proposes the establishment of internal Regional Consultative Committees as a precursor to the Regional Councils – very much in line with our own proposals arising from the AIR Project (see Symes and Phillipson, 1996). One important aim in devolving management from the functions of central administration is to bring together as wide a range of actors engaged in fisheries and their administration as possible. This approach would clearly assist the bottom-upwards flow of information, assessment and advice from the local and regional levels to inform the CFP as a whole. It would also help to promote a closer and functionally relevant relationship between the Council of Ministers, the Commission, the individual member state governments, the scientific community and the fishermen's organizations who would be given direct access to regional policy-formulating processes.

Second, where the 'regional seas' are shared between EU member states and other countries, the Regional Councils would form an important first point of contact for the joint negotiation of technical measures, access rights and resource allocation. And, third, there will be some instances where the regional seas, defined by reference to ICES divisions, will straddle both the EEZs of individual states and the high seas. This again creates an opportunity to ensure the coherent management of transboundary stocks, including the 'new' deep-sea fisheries located on the continental slopes either just within or just outside the 200-mile limits and which stand in special need of precautionary management (Hopper, 1995). Indeed, looking beyond the immediate concern for reform of the CFP, it is possible to envisage the 'regional seas' approach ultimately replacing the narrowly based coastal state management and the ineffective framework of voluntary international commissions in a cascading spatial arrangement of global ocean regions, 'regional seas', para-national marine space and local inshore fishing zones. But, for the present, the 'regional seas' approach does not intend to depart from the existing underlying principles of the CFP but rather to reinterpret and strengthen the notions of equal access, non-discrimination and relative stability within a more rational geographical framework and through a system of policy-making which respects the ideas of proximity, sensitivity and flexibility in the interpretation of the principles of sustainable resource management.

There will, of course, be many dissenting voices to such proposals, reflecting several different views. There is scepticism about the strength of common purpose that can be found among different resource user groups either based within the same locality or exploiting the same fishing area. The identity of purpose demonstrated in closing the ranks against a common enemy – the foreign fishermen or the faceless technocrat – is often illusory and unlikely to be replicated in transnational deliberations. Indeed, a major weakness in the argument which seeks to increase user participation in the policy process is the very cleavage of user group interests along both geographical and sectoral planes.

More fundamentally, a reluctance on the part of the administrators to admit the principle of regional self-determination into fisheries management may stem from: assumptions that it would require considerable rearrangement of administrative structures at increased cost to the exchequer; fears that a decentralization of policy-making could lead to disintegration into rival blocs which might challenge the central authority in Brussels and/or the member state; and premonitions that the system would prove unworkable. Sceptics might also argue, *inter alia*, that there is no plausible justification for the 'regional seas' approach; that carving up management into arbitrarily defined regions makes no more sense than the 'nationalization' of resources through EEZs; and that incongruent strategies followed by neighbouring Councils might create a sense of incoherence in the management of Europe's fisheries as a whole and heighten fears of discrimination.

Most of these criticisms, however, apply equally well to the existing CFP or to coastal state management. At worst, the fate of the regionally based consultative organizations, as outlined above, would be for them to be sidelined as yet another series of 'talking shops' which serve only to mask the fact that effective management remains firmly located at the centre. The way to avoid such criticisms is carefully to frame the statutory arrangements which define the relationship between the Regional Councils, the Commission and the Council of Ministers. The intention is not to replicate the role of the Commission at the level of the 'regional seas' but to take responsibility for the detailed drafting of management policy away from the Commission and relocate it at the level of the Regional Councils.

CONCLUSION

The CFP was designed to provide a solution to a major resource management problem, namely the rational and equitable exploitation of a common property resource under conditions of extreme overcrowding, overcapitalization and overfishing. It is the contention of this chapter that the persistence of a uniform, horizontal approach across a vast and diverse 'common pond' has blunted the instruments of management. The chapter has also argued the case for a more carefully coordinated approach to resource use and socio-economic development, that is for the integration of resource conservation, the development of fisheries-dependent regions and marine environmental management at a regional level. Notions of integrated planning and management are typically strong in idealism but notably weaker in practice; nonetheless, integrated regional management is the only feasible approach to the 'rational and responsible exploitation of resources' on a sustainable basis (Council, 1992), which must lie at the heart of any fisheries management. Without such an approach, the impact of measures introduced under a common fisheries policy will always appear arbitrary, prompting resistance from resource users and distortions in their implementation so that their end results may well differ significantly from original intentions. Regionalization of the CFP is therefore not simply an option; it is essential for the development of effective, well integrated and equitable management of the resource base and for the welfare of the communities and regions which depend upon it.

'Regional seas' management is thus posited as the strategic direction for the reform of the CFP. Its goals are (a) to encourage the adoption of more ecologically sensitive management strategies for marine resources, especially bearing in mind Gislason's (1994) advice that the objectives of fisheries management are shifting from the sustainable use of commercially valuable resources to the conservation of marine environmental quality; (b) to improve the flexibility of management responses in the face of increasing uncertainties affecting not only stock estimation but also the conditions of the markets; (c) to induce a greater transparency of decision-making in a partnership of science, administration and industry in order to generate greater commitment to and compliance with the regulation of the fishery; and (d) to develop a more direct link between fisheries management and the socio-economic effects of policy decisions and so resolve the disturbing functional discontinuities of the two main elements of policy – resource conservation and the development of fisheries-dependent regions.

Advocacy of a 'regional seas' approach implies a real change in the attitudes and mechanisms of policy-making, for it cuts right across the prevailing convention of centrally controlled, technocratically inspired policy and top-down delivery. The proposals will not find favour in a European Union which is still obsessed with centralized policy formulation and unprepared to grant a significant degree of autonomy in policy-making to the regions. They would, however, suit a more mature, confident and sensitive Union, willing to recognize in full the logic of spatial and professional proximity in formulating management strategies for the sustainable development of a complex natural resource.

NOTES

1. Devolved and Regional Management Systems for Fisheries (AIR-2CT93-1392), Draft Final Report, submitted to the Commission, January 1996.

BIBLIOGRAPHY

Commission of the European Communities (CEC) (1991) *Report from the Commission to the Council and European Parliament on the Common Fisheries Policy* (SEC (91) 2288, final, 18.12.91).

Commission of the European Communities (CEC) (1992) *Regional Socio-Economic Studies in the Fisheries Sector* (Directorate-General for Fisheries).

CFPRG (1996) *A Review of the Common Fisheries Policy Prepared for UK Fisheries Ministers by the CFP Review Group, Volume 1: Conclusions and Recommendations* (London: MAFF).

Committee of the Regions (COR) (1995a) *Opinion of the Committee of the Regions on the Revision of the Treaty on European Union* (Brussels).

Committee of the Regions (COR) (1995b) *Draft Opinion ... on the Regional Consequences of the Common Fisheries Policy* (CdR 318/95).

Council Regulation (EEC) No. 3760/92 of 20.12.1992: *Establishing a Community System for Fisheries and Aquaculture* (OJL 389, 31.12.1992).

Fishing News (1996).

Gislason, A. (1994) 'Ecosystem Effects of Fishing Activities in the North Sea', *Marine Pollution Bulletin*, 29: 520–7.

Hopper, A. (ed.) (1995) *Deep-Water Fisheries of the North Atlantic Oceanic Slope* (Dordrecht: Kluwer).

Jentoft, S. (1989) 'Fisheries Co-Management: Delegating Government Responsibility to Fishermen's Organizations', *Marine Policy*, 13: 137–54.

Jentoft, S. and McCay, B. (1995) 'User Participation in Fisheries Management: Lessons Drawn from International Experiences', *Marine Policy*, 19, 3: 227–46.

Ministry of Agriculture, Fisheries and Food (1996) *Fisheries Conservation Group: Summary of Recommendations* (London: MAFF).

National Federation of Fishermen's Organizations (1993) *Conservation – An Alternative Approach, Report on Alternative Proposals to the Government's Sea Fish (Conservation) Act 1992* (Grimsby: NFFO).

Nielsen, J. R. and Vedsmand, T. (1996) *The Role of Fishermen's Organisations in the Fisheries Management Decision Making Process*, paper presented to the Colloquium on the Politics of Fishing, Newcastle University, 17 September.

Phillipson, J. (1998) 'The Political Culture of Fisheries Management: an Anglo-Danish Comparison of User Participation' (this volume).

Symes, D. and Crean, K. (1993) *Regional Self-Management: Towards a Socially Responsible Fisheries Policy*, paper presented at the European Association of Fisheries Economists, Brussels.

Symes, D. and Phillipson, J. (1996) 'The Imperative of Institutional Reform: Alternative Models and the UK Fishing Industry', in *Proceedings of the VIIth Annual Conference of the European Association of Fisheries Economists*, Portsmouth, UK, 10–12 April 1995 (CEMARE: University of Portsmouth), pp. 229–44.

Wise, M. (1996) 'Regional Concepts in the Development of the Common Fisheries Policy: The Case of the Atlantic Arc', in Crean, K. and Symes, D. (eds), *Fisheries Management in Crisis* (Oxford: Fishing News Books), pp. 141–58.

13 Coastal State Management

Barrie Deas

INTRODUCTION

The National Federation of Fishermen's Organizations (NFFO) strongly advocates a move towards coastal state management as a viable and desirable alternative to the present Common Fisheries Policy (CFP). This chapter contends that the principle of equal access to a common resource lies at the heart of the CFP's shortcomings and contains an inbuilt imperative towards a highly centralized system of control. This conflicts with an increasingly widely recognized need for devolved management systems and greater involvement by fishermen in shaping the regime which controls their lives.

The chapter briefly outlines the specific aspects of the present CFP arrangements which have fuelled an emerging awareness within the UK that the prospect of a viable future for the British fishing industry is foreclosed by the principle of equal access; it goes on to discuss the triggers which have crystallized a generalized dissatisfaction with the present arrangements into a conviction that radical change is both necessary and possible.

Finally, it makes the case for a system of interlocking coastal state management regimes to replace the present CFP and discusses the political factors which will dictate the pace and direction of change.

THE COMMON FISHERIES POLICY: EQUAL ACCESS, CENTRALIZATION AND FAILURE

Fishing issues in Britain, with a few historically notable exceptions, have rarely been close to the top of the list of national priorities. This and the general enthusiasm for the European Common Market, is sufficient to explain how the CFP was signed in 1970, against fishing industry pleas, on the basis of the principle of equal access to a common resource. That from the outset, the application of the principle, undiluted, would be unacceptable, is reflected in the six- and 12-mile limits derogation and later, in 1983, the relative stability formula for national quota shares – both departures from, and qualifications to, the principle.

Integral to the equal access principle was the transfer of the essential levers of control over fisheries policy from member states to Brussels. The Commission henceforth took priority in initiating policy and negotiating with third countries, and the Council of Ministers, (in which the UK was one voice among many) henceforth shaped and ratified fisheries legislation. Member states' fisheries role was therefore limited to implementing and enforcing the rules made in Brussels. Once the principle of equal access was conceded it followed that the institutional arrangements for fisheries management must reflect and embody this fundamental approach. For this reason, a CFP based on equal access carries an inherent imperative towards centralization. It is an approach in which authority is concentrated and control systems are necessarily undifferentiated and therefore rigid. Fish do not adhere to territorial boundaries drawn on a map and this provided at least part of the formal rationale for the unique treatment of fish as a common rather than a national resource. But that such a centralized system of management is an effective, or the only, way to deal with the problems arising from shared stocks is a proposition that is now increasingly being challenged. It is historically the case that only one option was pursued at the time the CFP was signed, or since. The difficult years of negotiations which preceded (and resulted in) the 1983 agreement left member states exhausted with the fishing issue and disinclined to unpick any element of the agreement for fear that the whole set of arrangements would unravel. In any event, the stability which the CFP had promised was, immediately post-1983, untested. High fish prices and the availability of construction and modernization grants provided, for a short time, a veneer of prosperity which masked underlying weaknesses and supported an illusion that, if not perfect, the CFP provided the basis for a viable future. A review of the CFP in 1993, midpoint in the full term of the CFP, essentially endorsed the status quo, sidestepping any fundamental consideration of core questions.

From the 1983 agreement to here is only some 13 years but the subsequent period is such a catalogue of failure that it has brought into question the basic tenets of the CFP and has generated powerful forces of opposition within the UK and beyond. Four main areas of dissatisfaction have emerged: conservation, relative stability/flagships, uneven enforcement and policies towards overcapacity.

Conservation

In the final analysis the principal criterion for judgment of the CFP must be its conservation and resource policy and no one can be found to claim that this has been a roaring success. The Commission, member states, scientists and fishermen all hold different views as to who or what is responsible for the failure to halt the decline of important commercial stocks but there are few, if any, who dispute the fact that effective remedial action is necessary. Overcapacity, the absence of realistic technical conservation measures, uneven enforcement, industrial fishing, an over-rigid quota system which encourages discarding, the failure to secure the confidence and support of fishermen, all have played a part. The net effect of the failure to implement measures which are simultaneously effective and at least tacitly supported by the fishing industry has been a spiral of declining spawning stock biomass and decreasing TACs.

Relative stability

The centrepiece and cement of the 1983 agreement was the principle of relative stability. Increasingly, the formal status of the principle has been emphasized while its substance has been undermined by quota hoppers. The contradiction between the quota shares agreed in 1983 on the basis of nationality and the Treaty provisions on the freedom of movement of labour and capital and the right of establishment, has become increasingly apparent. The degree to which vessels of one member state would obtain access to the register of another member state, with the express purpose of obtaining access to that member state's fish quotas, was wholly underestimated in 1983 but sanctioned and given further impetus by the 1990 European Court judgment in the 'Factortame' case. But the fact that 46 per cent of the UK's hake quota, 44 per cent of its plaice quota, 35 per cent of its megrim quota, 29 per cent of its angler fish and 18 per cent of the UK sole quota are now taken by quota hoppers is testimony to the extent to which relative stability has been and continues to be undermined.

Enforcement

British fishing industry claims that the CFP is enforced on them but only to an uneven degree in many other member states have often been met with a degree of scepticism. It has been said more than once 'well they would say that, wouldn't they?' But the Commission's own report on enforcement in the CFP, published in 1996, spells out the discrepancies. With reference to the UK it says:

> The United Kingdom has a well developed national fishery control system, which is matched by the allocation of substantial resources.

> The Commission acknowledges the efficient and competent manner with which the United Kingdom has tackled the enforcement task at sea.

> The means and the commitment invested in control and inspection in the UK provide an example of how the CFP should be enforced.

While of other member state it records:

> The organization of monitoring, control and surveillance differs considerably from one Member State to another.

> This results in a broad range of organization types, varying from a comparatively well-organized service using qualified staff in one Member State to a poorly coordinated set of national and regional departments with non-specialized personnel in another Member State.

> The number of inspectors in each port differs substantially from one Member State to another. For some countries there are several inspectors in each fishing port, whilst in others one fisheries inspector is in charge of several ports. The low level of human resources in some Member States raises doubts as to whether the control regime applicable to the common fisheries policy is applied there. Several Member States have insufficient specialized equipment to meet their monitoring obligations, even if some of them have used Community financing to upgrade their resources during the past five years. A small number of Member States even have no airborne surveillance, which considerably reduces the efficiency of inspections at sea.

There are large differences in the approach to and effectiveness of data collection and verification – even though the official reports always try to present the existing systems in a positive light. It is also evident that all Member States could benefit from the experience of other Member States in refining their own systems.

(CEC, 1996)

This simply confirms the abundant anecdotal evidence that, after 12 years, there is far from full compliance with the CFP and that UK fishermen are subject to what is by far the most stringent enforcement regime of any EU member state. They legitimately ask what is 'common' about such a system. Part of the Commission's reaction to the evident failure of the CFP's resources policy has been to advance additional and tighter control measures. The net result has been the development of a complex system of rules, difficult to comprehend and in any event unevenly enforced, as the Commission's report confirmed.

Overcapacity

It is a bald but unpalatable fact that there is a severe imbalance between fishing capacity and available resources. In recognizing this fact, the main UK industry organizations have supported measures taken by the UK government to reduce the size of the fleet, in compliance with the EU's third Multi-Annual Guidance Programme (MAGP). After a long delay and a false start, the UK government implemented a stand-alone, voluntary decommissioning scheme in 1993.

Proposals for a fourth generation Multi-Annual Guidance Programme are being discussed at the moment and the UK's view is heavily coloured by the fact that some 20 per cent of the tonnage of the UK fleet is now in the Spanish or Dutch quota hoppers' hands. In effect, these member states have 'exported' their overcapacity problem to the UK while simultaneously enjoying the economic benefits of UK quotas. There are also many fundamental flaws in the proposals for MAGP IV, not least the lack of coherent scientific justification for the levels of cuts proposed for different fleet segments and the fact that the relative age and efficiency of different national fleets is simply not taken into account. The present direction of the EU fleet reduction programme is towards the replacement of a large obsolete fleet by a small efficient fleet, with significantly greater fish-killing power. The conservation implications of this do not need spelling out but are so far unaddressed in the Commission's proposals.

Because the UK Treasury blocked a British decommissioning scheme until 1993, the UK failed to meet its fleet reduction targets for MAGP III; consequently, construction and modernization grants have been denied to British fishermen at the same time as other member states, notably Spain, have seen substantial renewal. Not unexpectedly, the UK fleet is ageing – it now has an average vessel age in excess of 25 years – which carries serious safety implications, in addition to the obvious competitive disadvantage. Hence, on this occasion with UK government complicity, the CFP has overseen the decline of the fabric of the UK fleet while the Dutch, Spanish, French, Danish and Irish fleets have developed and modernized.

I catalogue these failures and inequities within the CFP because collectively they underpin the deep sense of frustration and bewilderment evident throughout the fishing ports in the UK. An increasingly dispirited and alienated industry has sought to make sense of the morass of rules which govern their every activity at sea but which are clearly ineffective in their central objectives of stock conservation and of providing a broadly equitable management system. Add to this the patent unfairness of the uneven enforcement of the rules in other member states, and the perversity of flagships taking quotas explicitly intended for the exclusive use of British fishermen, and it is not necessary to turn to xenophobia to explain fishermen's profound disillusionment with the CFP. It is a quite natural and understandable reaction to an invidious and wholly unacceptable situation not of their own making.

The failure of the CFP at root is not essentially one of institutional rigidity or even absence of political will, although these elements are present. At base the CFP fails because it faces a crisis in legitimacy. It is not accepted as an effective and equitable way of managing our fish resources. Based on the principle of equal access it can only govern from the top down on proposals from the Commission; by inherent design it cannot be responsive to the contours of the industry in different parts of the Community. Blanket measures, well intentioned perhaps, but crude and inflexible in application, have been the hallmark of the CFP. And the result has been increased distance between the rulers and the ruled, to a point where the legitimacy of the former is challenged by the latter.

Paradoxically, it was the UK industry's vigorous opposition in 1992–4 to the British government's proposals for a days-at-sea regime, through a highly effective mass rally, port blockades and finally legal action, which demonstrated that unacceptable measures could be successfully resisted; change could be brought to the management regime and continual reverse and slow decline need not be the inevitable fate of the UK fishing industry.

However, before the industry's demand for change settled on the basic underpinning of the CFP, a trigger was necessary and this was provided by the Commission's proposals for a highly bureaucratic effort control regime in western waters.

WESTERN WATERS ARRANGEMENTS

It was the Western Waters Agreement of December 1994 which convinced many in the UK industry that damage limitation was not by itself an adequate response from a beleaguered industry. The Western Waters Agreement in many ways can be regarded as a turning point in that it convinced many within the industry that a radical new approach to the CFP was necessary. The sight of another British minister returning from the Fisheries Council in Brussels yet again to claim, truthfully, that this was the best deal possible in difficult circumstances was the straw that broke the camel's back and triggered a reaction which is still working its way through.

Given the Spanish fleet's record of targeting immature fish, the decision to permit access of 40 Spanish vessels into the Irish Box, an area previously closed to them, was of course provocative and elicited a strong reaction. The imposition of a further layer of reporting requirements, as part of a system of effort control in western waters, further stoked resentment from an industry already suffocating with bureaucracy. However, it was the fact that to facilitate the potential entry of Norway into the European Union, full Spanish and Portuguese accession to the CFP was brought forward a full six years, setting aside previously agreed and legally enshrined arrangements, that provided the central lesson learned by the UK industry from the Western Waters Agreement. The message received and understood was that, notwithstanding the niceties of the European legal and political framework, set and binding agreements can be overturned, in the right circumstances, by the application of political pressure. The Spanish government had secured these changes on behalf of its fishing industry; the UK government, with political will, could do the same.

The Save Britain's Fish campaign which had been in existence, if not at the forefront, of fishing politics for some years was growing in strength and took on a new impetus from the Western Waters Agreement as the principal focus for dissatisfaction with the CFP. The campaign acted as a catalyst, transforming a generalized dissatisfaction into a more focused and purposeful movement for change. Although the SBF campaign has publicly been closely associated with the Eurosceptics, it draws wide support both from those who are pro-Europe but anti-CFP, as well as from those who hold a broader anti-European political agenda. The common denominator is dissatisfaction with the present state of affairs and the conviction that the principle of equal access underpins many of the industry's ills.

COASTAL STATE MANAGEMENT

The general international acceptance, during the course of the 1970s, of fisheries jurisdiction out to the 200-mile limit, provides the geopolitical underpinning for the NFFO's view that coastal state management is an appropriate successor to the present CFP. Britain's 200-mile limit was absorbed into the EU 'pond' in 1976 before it could come into effect in its own right and of course Britain, in acceding to the CFP, jettisoned its right to apply its own management regime within its 200-mile fisheries limit. That this, in the NFFO's view, was a fundamental wrong turning has amply been confirmed by the failure of the CFP in the intervening period. Having arrived at the conclusion that the CFP, based on equal access, must go, it became a pressing necessity for the Federation to develop a coherent, viable, alternative model and that is the work in which we are presently engaged. It is not, however, possible to provide a detailed blueprint of how a coastal state management regime would operate. Broad themes, concepts and principles are available but in practice much will depend on the circumstances in which such a policy is implemented and the bilateral arrangements agreed with other countries with whom we would share stocks and reciprocal agreements. Fundamentally, however, it is envisaged that, following a break with the principle of equal access (in whatever way achieved), EU member states would be in a position to apply and enforce their own rules of access, their own technical conservation measures and their own structural framework under a system of interlocking coastal state management regimes. This may take place under revised CFP arrangements or following a more profound break with Brussels, but the end result would be to devolve management responsibility to the member state. The subjection of fisheries policies to the lowest common denominator and to qualified majority voting would end and member states would be freed to pursue conservation policies which would give their fishing industries some assurance of a viable future.

In broad and schematic outline form the framework for coastal state management in the UK which we advocate takes the form shown in Figure 13.1.

Figure 13.1 provides the schematic form but we look to Norway as providing an operational model of a European country, with important fish resources under its own control, operating a coastal state management regime. Without suggesting that the Norwegians have done everything right, or that everything the Norwegians do is appropriate for our fisheries, it does provide an example, close to home, of how such a system can operate in practice. The difference between the success of Norway in rebuilding its stocks and the multiple failures of the CFP is too stark to ignore.

It is a truism noted earlier that fish do not adhere to political boundaries and therefore a system of interlocking coastal management regimes would, of necessity, require a series of bilateral agreements, both to allow reciprocal access arrangements and to establish a common approach to joint stocks. However, both in concept and practice, this is far from the heavily centralized, unresponsive and bureaucratic option offered under the equal access principle. Equal access predetermines the long-term direction of policy towards centralization. By contrast, coastal state management is predicated on the devolution of authority to the member state and beyond. In this it is at one with many practical trends and academic findings which suggest that bottom-up policies, which harness the views and involvement of fishermen in making and applying conservation policies, are more likely to achieve their conservation objectives than rigid and inflexible policies imposed from above.

Perhaps the CFP can be reformed to become more effective in conservation, less bureaucratic, more flexible, more equitable and more responsive to the industry's concerns and initiatives. But to move in this direction without a break from the principle of equal

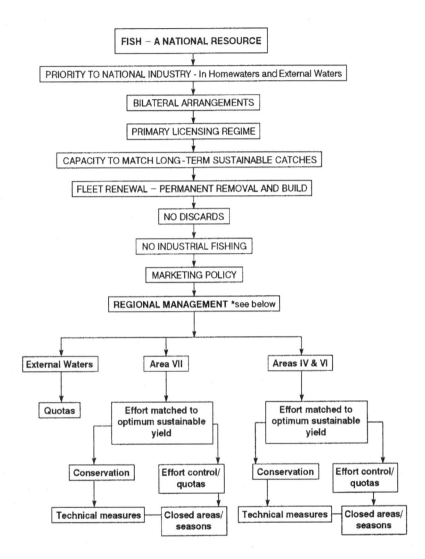

Figure 13.1 Framework for coastal state management in the UK.
* Confines management responsibility to those prosecuting the fisheries
in each region.

access would be to move against the grain of present institutional arrangements. Decentralization without a concomitant transfer of authority to member states is always likely to be a half-hearted compromise, more cosmetic than real. For this reason the 'regionalization' of the CFP would be a welcome but inherently inadequate development. Restricting decision-making to those member states with existing quota rights in a zone such as the North Sea or Irish Sea, clearly makes sense and is certainly preferable to the present situation in the Council of Ministers where Greece has a vote in North Sea fisheries issues, the UK has a say in the regime which applies in the Mediterranean and qualified majority voting applies throughout. But to implement such arrangements beyond the merely consultative would require a basic jurisdictional change at the heart of the CFP which would potentially destabilize the core principles of the CFP.

In any event, against this one qualified and limited proposal for decentralizing, the CFP has to be set several other current proposals which, if implemented, would mark a further significant deepening in its centralization:

- proposals for technical conservation measures which simplify the technical conservation rules for the legislators and administrators but which, at the level of the fishing vessel, remove operational options and increase the complexity of enforcement;
- fleet reduction programmes under MAGP IV which, if implemented on the Commission's proposals, would impose rigid fleet categories to prevent diversification by area or gear type when this type of diversification has been a crucial element in maintaining vessel viability and developing new fisheries;
- new control regulation proposals will add to the reporting requirements in logbooks and apply unrealistic tolerances on catch estimates, all driven by the need to improve the flow of information to the centre;
- proposals for a satellite monitoring system – the epitome of centralized control – to be followed in due course by electronic logbooks will further reinforce the 'Big Brother' syndrome;
- Commission progress towards an EU licensing system is a necessary precondition for completely centralized control from Brussels.

In its legal basis and its institutional form, the CFP is predicated on a centralized, uniform system and the Commission's recent policy initiatives fully reflect that approach. It is a top-down approach, of necessity inherently antipathetic towards initiatives from below.

There is, therefore, a clear and stark political choice facing us. Although a senior Commission official observed recently that 'the CFP is comprised of a number of objectives none of which are compatible with each other', the principle of equal access ensures that the CFP will continue to move, inexorably, towards the centralized option. While that principle remains, any movement in the opposite direction will be inadequate and largely cosmetic. By contrast, the NFFO and its allies, within and beyond the UK, argue for a break with the principle of equal access and for the introduction of coastal state management as a workable and effective alternative. Such a framework would provide the institutional basis for a decentralized system of fisheries management in which decision-making could genuinely be devolved. Devolved management can take many forms and the UK's system of quota management in which producer organizations manage annual sectoral allocations is a pointer in the direction we wish to travel. It is, however, only a pointer and the institutional apparatus to go further is currently absent. Fishermen have extremely limited influence in shaping the rules which apply to their fisheries.

An example of what could potentially be done in this direction was seen during 1992 in the context of the UK government's attempts to impose a days-at-sea regime on the UK industry. Against this background, the NFFO developed an alternative set of conservation proposals based on technical measures centred on improvements in fishing gear selectivity. The approach adopted was to convene a series of meetings in which fishermen working a particular type of gear in a particular area for particular groups of target species would agree measures which they would accept. Ministry scientists participated, evaluated the effectiveness of each proposal and commented generally. The result was a document *Conservation: An Alternative Approach* (NFFO, 1993) which if implemented could form the basis of an effective technical conservation regime. More than its immediate relevance, however, is that the process through which the proposals were developed is, potentially, a model of how rules, accepted and agreed by the industry, could be formulated by and for the industry. A bottom-up approach has a degree of legitimacy which a top-down approach can never have – in addition to the advantage of framing rules sensitive to the real contours of the industry.

But coastal state management is not a panacea. The subsequent stock collapses of northern cod stocks off eastern Canada and the Icelandic cod fisheries – both under the control of a single nation state – auger against the over-optimism and over-expansion which can accompany the euphoria of obtaining exclusive control. Coastal state management can provide the framework but the policies within that framework must be effective and realistic. And we are well advised to remember that, although developed in response to EU fleet reduction targets, the 1992 days-at-sea regime was a home-grown bureaucratic nightmare.

Coastal state management would allow the UK's share of the resource to more accurately and fairly reflect its contribution to the resource (the 1975–9 reference period baselines being seriously flawed in many respects). However, restricted entry licensing would, of necessity, continue to lie at the heart of the management system. Equally, it is recognized that the fishing capacity must be adjusted to the size of available resources; appropriate technical measures must apply and fleet restructuring must ensure that the fleet is modern, competitive and safe, as well as in alignment with the biological base.

Every system, no matter how much an improvement on its predecessor, will face its own inherent limitations and drawbacks and will be introduced under a concrete set of political circumstances. Because a movement to a coastal state management regime does involve a number of important unknowns, the NFFO commissioned Hull University to undertake a study and prepare a report (HIFI, forthcoming) to present an impartial assessment of the costs as well as the benefits of such an approach. This report is now being assessed and will inform future NFFO policy in this area.

CONCLUSION

Policy initiatives do not take place in a vacuum and the receptiveness of governments, politicians and the public to the radical changes we propose is influenced by many factors outside the industry's control. Tides of fortune flow and ebb and there is no doubt that the British government's current sensitivity to the fishing industry's concerns is not unrelated to its slender parliamentary majority and vulnerability on the European issue generally. This leverage has moved fishing further up the political agenda than it has been for a generation and the prominence of the flagship issue in the UK's list of demands at the Intergovernmental Conference is testimony to its current significance.

The government is right to focus initially on the flagship issue because it is the single most damaging and inflammatory aspect of the CFP, undermining expressly agreed and understood intentions of the 1983 agreement. Resolving the flagship issue is therefore critical for any future fisheries management regime, whether based on the principle of equal access or not, because the issues involved derive from the Treaty of Rome. Resolution is essential but shoring up relative stability, important as that is as an interim policy, is insufficient by itself. A more radical and far-reaching approach which goes to the heart of the present CFP is demanded not just by British fishermen but by a dispassionate analysis of the patent failings of the current arrangements.

A large majority for either of the major parties in the next general election would remove important leverage from the fishing industry but would not mean that the problem and therefore the issue would go away. The strength of the European Union to withstand broad nationalist pressures, the extent to which the Commission will be brought under real democratic control and the extent to which the CFP can reform itself, as well as the pattern of domestic politics, all potentially shape the circumstances in which a coastal state management regime might be introduced. It is probably pointless to speculate further, beyond the observation that the fracture lines in British and European politics will be decisive in determining the pace and direction of change.

Pressure for change from the industry will most certainly continue and it is significant that both the UK government and opposition parties have, in their own ways, acknowledged the need for greater exclusivity of control within UK waters. The Conservative government, at least partially in response to industry pressure, is at the forefront of those pushing for regional management as a way of blocking potential demands for access arising from the expansion of Europe and the accession of new member states; the Labour Party has voiced its anxieties over the CFP's basis in the concept of a common resource, when such a principle is clearly incompatible with the welfare of the British fishing industry; the Liberal Democrats have no difficulty in supporting the concept of radical reform of the CFP. Eurosceptics have of course a broader agenda in Europe but have used Parliament to great effect to highlight the significance of the fishing issue.

There is now, therefore, notwithstanding the main UK political parties' commitment to Europe, a significant and growing body of political support for movement towards coastal state management as an alternative to the present CFP. Both the political will to force through the necessary changes and the coherence and credibility of the coastal state management model will be the principal focus of the NFFO's efforts in coming months.

BIBLIOGRAPHY

CEC (1996) *Monitoring the Common Fisheries Policy*, Commission of the European Communities, COM (96) 100 Final.
NFFO (1993) *Conservation: An Alternative Approach* (Grimsby: NFFO).
HIFI (forthcoming) *Coastal State Management* (Hull: Hull International Fisheries Institute).

14 The Birth of the Aquarium: the Political Ecology of Icelandic Fishing[1]

Gísli Pálsson

INTRODUCTION

Focusing on Iceland, this chapter both compares the ecological knowledge of fishers and professional marine biologists and examines changes in their relative importance in public discourse and resource management in the course of recent history. My approach is informed by both the theory of practice, with its notions of situated action emphasizing the role of direct engagement with everyday tasks, and Foucault's conception of social discourse. The empirical data used are based on written documents as well as interviews and ethnographic fieldwork.

Long ago, Marx argued that in the case of 'extractive industries' such as fishing, the 'material for labour is provided directly by nature' whereas other modes of production deal with 'an object of labour which has already been filtered through labour' (1976: 287). Such a notion was, no doubt, reinforced by the widespread Western idea that the supply of living resources in the ocean was a boundless one. Thus Thomas Huxley wrote in 1883: 'I believe that the cod fishery . . . and probably all the great sea-fisheries are inexhaustible; that is to say that nothing we can do seriously affects the number of fish' (cited in McGoodwin, 1990: 66). Neither position, of course, is tenable in the modern world. Many of the world's major fishing stocks are threatened with both over-fishing and pollution – oil, radioactive waste and other by-products of human activities – and fisheries more and more resemble other branches of industry in that the resource base is increasingly subject to deliberate human impact. For one thing, while the structure and size of extensively migratory fish populations are continually subject to

209

extreme uncertainties, the boundaries of 'wild', extractive fisheries
are increasingly becoming blurred, with exponential growth in sea
ranching and fish farming and their growing impact on 'natural'
stocks. Consequently, to think of the oceans as a boundless store-
house of living resources unaffected by humans – as a source of
'material for labour . . . provided directly by nature', in Marx's ter-
minology – really does not make much sense. It would be far more
appropriate to speak of an aquarium. Current fisheries management, I argue, based on the allocation of individual property rights
to fishing stocks or shares in 'total allowable catch' (TAC) for a
season, exemplifies what Gudeman has identified as the 'modernist
production regime' (1992: 151), a regime based on the idea 'that
the human and natural world can be organized and subjected to
rational, totalizing control'. At the same time it represents the cul-
mination of a complex cognitive and political history that may be
summed up as the birth of a gigantic aquarium – to paraphrase
Foucault about the 'mutation in discourse' embodied in the medical clinic (1973: xi).

Focusing on Iceland, this chapter examines changes in the relative importance of the ecological knowledge of fishers and professional marine biologists in public discourse. I shall examine some
aspects of the discursive changes in Icelandic fishing which led to
scientific management and state control, by a rather cursory analysis of written contributions of fishers, managers and scientists to
the journal of the Icelandic Fisheries Association (*Ægir*), published
since 1905. While modernist management tends to ignore the potential of the practical knowledge of fishers, there are good grounds,
I suggest, given the apparent uncertainty and chaotic nature of many
marine ecosystems, for exploring how practical knowledge is acquired, what it entails, and to what extent it could be brought more
systematically into the process of resource management. On the
other hand, the attempt to collect and store 'indigenous' knowledge is seriously misguided. I shall critically examine one collaborative attempt involving both fishers and professional marine
biologists – the Icelandic 'trawling rally', the repeated trawling, under
controlled conditions, along predetermined paths. My approach is
informed by both the theory of practice, with its notions of situated action emphasizing the role of direct engagement with everyday tasks, and Foucault's conception of social discourse. The empirical
data used are based on written documents as well as interviews
and ethnographic fieldwork.

With Iceland's independence in 1944, fishing effort on Icelandic fishing grounds multiplied. In 1948 the Icelandic Parliament passed laws about the 'scientific protection of the fishing grounds in the coastal zone', to be able to prevent over-fishing of its major fishing stocks, particularly cod. Four years later, Iceland announced that it would extend its territorial jurisdiction from three to four miles and in 1958 it unilaterally extended its jurisdiction to 12 miles. In 1976, the Icelandic government extended the national fishing limits to 200 miles which marked the end of the last Cod War with Britain and West Germany. The domestic fishing fleet, however, continued to grow and catches, relative to effort, continued to decline. The first serious limitations on the fishing effort of Icelandic boats were temporary bans on fishing on particular grounds. By 1982 politicians and interest groups were increasingly of the opinion that more radical measures would be needed to limit effort and prevent the 'collapse' of the cod stock. A boat-quota system was suggested in 1983 to deal with the ecological and economic problems of the fisheries, a system that would divide this reduced catch within the industry itself. The precise allocation of catches was debated, until it was agreed late in 1983 that each boat was to be allocated an annual quota on the basis of its average catch over the past three years. This meant that some boats would get higher quotas than the rest of the fleet, a fundamental departure from traditional policy (see Pálsson and Helgason, 1995).

Early in this century, fishers and boat owners gradually became the central agents of production discourse, replacing the landed elite as the economy shifted from stagnant agriculture to expansive fishing. Now, once again, with scientific management and a quota system in the fisheries, the discursive pendulum has swung in the opposite direction – from sea to land. Fishing remains a major economic enterprise, but the makers of knowledge and economic value are no longer fishers but the land-based owners of boats and fishing plants and the holders of scientific, textual knowledge.

Related to the regime of the aquarium is the idea that scientists and practitioners represent fundamentally different cognitive worlds, the former belonging to the outside world, the latter being inside, along with whales and fish and other inhabitants of the aquarium. Such an idea, of course, is not restricted to fisheries. For several centuries, since the Renaissance and the Enlightenment, Western discourse has tended to radically separate scientific understanding and everyday accounts. Scientists, it has often been assumed, are objective explorers of reality, proceeding by universal concepts, rational methods and detached observations, while the lay person is locked up in a particular world, driven by genetic make-up, ecological context and local concerns. One of the consequences of such a scheme is the tendency to reduce local environmental knowledge to mere trivia. Accordingly, sustainable resource use and sensible management become the privileged business of outsiders formally trained in public institutions.

SHIFTING DISCOURSES

Icelandic production discourse has undergone a series of successive changes as Icelanders have assumed new kinds of relations in the course of appropriating marine resources (Pálsson 1991). To each phase in the development of Icelandic society corresponded a particular dominant 'paradigm', an underlying framework of understandings and assumptions with respect to what constitutes production and ecological expertise. One of the important changes concerns the discursive shift from land to sea at the turn of the last century. During earlier times, Icelandic farmers and landowners occupied a central position and, consequently, fishing was regarded as merely a supplementary subsistence activity. Fishing was not just a marginal occupation; it was also the subject of a cultural struggle. This can be seen from the fact that in the nineteenth and early twentieth centuries those in power tended to present fishing communities as 'devoid of culture' (*menningarsnauð*), the source of degeneration, alienation and deficient language. Finnur Jónsson, an Icelandic ethnologist, remarked in 1945 with respect to a fishing village on the south-west coast that while it was always regarded 'as one of the best fishing places in the south, . . . its culture was at a rather low stage of development' (quoted in Pálsson and Helgason, 1996b). In the nineteenth century, however, as new markets for Icelandic fish were developed, especially in Spain and England, fishing villages grew and there emerged an expanding market economy. While for many Icelanders the agricultural community continued to provide the dominant cultural framework, the 'essence' of the Icelandic way of life, the focus of discussions on economics and production inevitably shifted from the landed elite to the grass roots of the fishing communities as fishing became a full-time occupation and a separate economic activity. Fishers, particularly skippers, became the key figures of production discourse. In the process, agriculture was redefined as a burden to the national economy.

In Iceland, some marine biological research had already occurred by the end of the nineteenth century, but full-time research started later, in the 1940s. Most of the contributions to the first issues of the journal of the Fisheries Association are those of fishers, but gradually marine biologists enter the scene. With fishers being important agents in the expanding economy, marine scientists had to carve a space for themselves in the role of collaborators and apprentices. This is especially clear in the writings of Bjarni Sæmundsson, the pioneering ichthyologist. In the 1890s he travelled for several years throughout Icelandic fishing communities to learn from practising fishers:

> I had the opportunity to observe various kinds of newly-caught fish, to look at fishing gear and boats and to listen to the views of fishermen on various matters relating to fishing and the . . . behaviour of fish.   (*Ægir*, 1921, 14: 115)

Sæmundsson seems to have thought of himself as a 'mediator' (*millili∂ur*) between foreign scientists and Icelandic fishers (*Ægir*, 1921, 14: 116), eager to learn from both groups. Scientific knowledge, along with the 'practical knowledge' (*reynsluþekking*) of fishers, he suggested, was 'the best foundation for . . . the future marine biology of Iceland' (*Ægir* 1928, 21: 102). The pioneering biologists did not only regard themselves as humble apprentices, they were moderately optimistic about the immediate achievements of the scientific enterprise. Referring to the prospects of dealing with 'the old mystery, the migration of fish', Sæmundsson comments:

> We should not expect . . . to be able to deal with everything and, thus, to answer whatever question we may have, for instance to establish the location of herring . . . at a particular time. That kind of knowledge is far away, although it is our mission (*hugsjón*) to be able to provide it in the future.   (*Ægir*, 1924, 17: 144)

Such a mixture of mission and modesty was, no doubt, necessary in the beginning, to provide political and financial support for marine science. Gradually, the necessary trust and confidence were attained; fishers, boat owners and the general public participated in establishing fisheries science. Significantly, in 1931 one of the regional Fisheries Associations (which embrace most interest groups) resolved:

The behaviour of most of the fish species we exploit are now known in most respects. The place and time of spawning . . . are topics which science has for the most part mastered. Nevertheless, there are many 'dead spots' in our knowledge of fish behaviour. Little or nothing is known about what determines fluctuations in the catch.   (*Ægir*, 1931, 24: 29)

At the same time, however, another discourse emerged which downgraded the expertise of fishers. While the editor of the fisheries journal urged fishermen to participate in discussions on the fisheries, he sometimes reinvented the biases of the earlier agricultural elite:

> You fishermen! This journal is intended for you, it should be your guide and your voice . . . it should speak for you when you are busy at sea . . . it should enlighten those of you who live at the outskirts, at the margin where the profit is, but often, too, unfortunately, ignorance and poverty.   (*Ægir*, 1932, 25: 159)

By the middle of the century, the subtle competition of the discourses of fishers and biologists seems to have developed into open confrontation. In 1947 the fisheries journal published an article which found it necessary to remind the readers of the journal of the significance of practical knowledge:

> Fisheries research is . . . intimately connected to fishing . . . and naturally . . . it should be carried out in collaboration with perceptive fishermen and boat owners . . . *This may not be particularly scientific, but we should keep in mind that the experience that perceptive fishermen have acquired after years of practice . . . must provide some kind of guidance to the scientists. It is no coincidence that the same men catch more than others year after year.* What matters most is their attentiveness and their perceptiveness with respect to the behaviour of cod and herring. Icelandic ichthyologists should recall Bjarni Sæmundsson who once remarked, in his well-known humble spirit, that he owed most to the fishermen of this country . . .
> (*Ægir*, 1947, 40: 159, emphasis added)

During the cod wars, the biologists at the Marine Research Institute, established in 1965, emphasized the prospects of estimating the composition and size of the cod stock:

In recent years, the success of spawning for a given year has been extensively studied. We have obtained tentative estimates of the size of different year classes, but since such research only began recently . . . it is not quite clear . . . what each year class will supply for the fisheries. Before long, however, we should know, and then we expect to be able to predict, only a few months after spawning, the real size of the year classes. I am optimistic that in the future we will be able to make forecasts with more accuracy than at present.   (Schopka, 1975: 48)

Now, two decades later, these words sound overly optimistic. Fundamental ecological relationships – including the relationships between the size of the spawning stock, the success of spawning and the size of the future fishing stock – have turned out to be far more complex and difficult to establish than the biologists estimated.

The tone of humility and mutual learning typical of the pioneering biologists during the first half of the century has been replaced by claims about scientific certainty and folk 'misunderstanding' (see *Ægir*, 1964, 57: 109). The element of trust that characterized relations between scientists and fishermen has evaporated. One skipper provided the following observation:

When fisheries biologists realize that fishermen possess knowledge which they themselves do not have, and when these two groups begin to co-operate on the basis of each other's knowledge, then we may envisage realistic knowledge about the quantity and behaviour of cod on Icelandic grounds.

(Guðjón Kristjánsson, *Ægir*, 1979, 72: 595)

Another skipper remarked a few years later: 'Fisheries biologists have no possibility of finding or counting the fish in the sea. They have no equipment for this purpose beyond those that fishermen have' (Guðjón Sigtryggsson, *Ægir*, 1986, 79: 33).

Why did this shift in discourse take place? For one thing, during the cod wars, marine science played an important role. Scientists supplied the arguments necessary for the 'rational' management of the fisheries. In the words of a former Director of the Fisheries Research Institute: 'Our rights will never be accepted internationally unless we are able to supply solid evidence, based on scientific research, for the dangers posed to our fisheries and our whole cultural existence' (*Ægir*, 1947, 40: 258). And with the birth of the modernist fisheries regime and the grand narrative of marine science, the voice of fishermen was gradually subdued, if not silenced. Another reason may have to do with shifting paradigms in marine biological research.

Marine sciences restrict the scope of fishing operations, in particular setting the limit of the total allowable catch (TAC) for each different species during a fishing season on the basis of their measurements and estimates, but it is primarily the science of resource economics that provides the theoretical rationale for the current economic management of fisheries. Economists commonly suggest, with reference to the 'tragedy of the commons', that over-fishing is inevitable as long as the fishing grounds are defined as 'common property', i.e. where access is free for a large group of producers, and that the only realistic alternative – euphemistically defined as 'rights-based' fishing (see Macinko, 1993: 946) – is a quota system. By instituting private property rights to the fishing stocks in the form of quotas and letting the market regulate their distribution, it is argued, rational production will theoretically be attained; assuming a sense of responsibility among the new 'owners' of the resource (the quota holders) and an unhindered transfer of quotas from less to more efficient producers, privatization both encourages ecological stewardship and ensures maximum productive efficiency (see Scott, 1989: 33). In the Icelandic context, quotas were originally attached to boats and there were severe restrictions on their exchange and transfer, but with the laws of 1990 quotas were separated from boats, becoming fully transferable. According to Árnason, an Icelandic fisheries economist, such a *gradual* transition to the full privatization of marine resources was unavoidable to alleviate opposition based on 'traditional values and vested interests rather than rational arguments' (1993: 206).

While the present management regime represents the apex of modernist management, with science in the leading role, there is a growing unease about the rhetorical context and the political role of science. Significantly, the Boat Owners' Association has hired a marine biologist on a permanent basis, to re-analyse the data of the Marine Research Institute and to facilitate a better understanding among politicians and the general public of the concerns of boat owners. This decision, apparently on the advice of boat owners in Canada, was triggered by discontent with the prognoses of the Institute and its conservative recommendations regarding the total allowable catch. Also, the results and recommendations of both the Marine Research Institute and the economists responsible for the establishment and design of the ITQ system have increasingly been challenged from within the scientific community, notably by researchers at the University of Iceland.

PRACTICAL SKILLS

It is increasingly apparent to many scholars that the local view of those directly engaged with everyday tasks makes much more sense than supposedly objective 'observers' have often assumed. This is evident from current interest in practical knowledge in development agencies on the international scene as well as in academic studies of learning and expertise. In recent years, development experts and resource managers have drawn attention to what is commonly referred to in the literature as 'indigenous' or 'traditional' knowledge. Some have focused on the need to maintain or 'recapture' the disappearing knowledge of marginal groups (see, for instance, Chapin, 1994), others have emphasized the potential importance of such knowledge for sustainable resource use (DeWalt, 1994), and still others have emphasized the need to defend the intellectual property rights of the groups they have studied against the hegemonic practices of states and multinational companies, focusing on legal claims about patents and royalties (Brush, 1993). At the same time there has been a growing theoretical interest in 'outdoor' learning (Lave, 1988; Pálsson, 1994), cognition 'in the wild' (Hutchins, 1995), and the divide between practical and scientific knowledge (Agrawal, 1995, Latour, 1994). The community of modellers, then, has been both expanded and redefined, empowering the local voice and relaxing modernist assumptions of privilege and hierarchy (Gudeman and Rivera, 1990). It is not quite clear, on the other hand, what the empowering of the local voice entails.

One of the important issues involved concerns the concepts of 'indigenous' and 'traditional' knowledge. While it is true that an extensive body of local knowledge has often been set aside, if not eliminated, in the course of Western expansion and domination and there are good grounds for attempting to recapture and preserve what remains of such knowledge (Chapin, 1994), the reference to the 'indigenous' and 'traditional' in such contexts tends to reproduce and reinforce the boundaries of the colonial world, much like earlier notions of the 'native' and the 'primitive'. How old does a particular skill or body of knowledge have to be to count as 'traditional'? Such terms are not only loaded with the value terms of colonial discourse, they are fraught with ambiguity (Agrawal, 1995: 415).

Another contested issue relates to the meaning of knowledge and learning. Orthodox theories tend to present the learning process in highly functional terms, presupposing a natural novice who gradually becomes a member of society by assimilating its cultural heritage. Knowledge becomes analogous to grammar or dictionaries, invested with the structural properties and the stability often attributed to language. Given such a perspective, indigenous knowledge is sometimes presented as a marketable commodity – at times with 'missionary fervour' (DeWalt, 1994: 123). It may be useful and quite legitimate in some contexts to think of practical knowledge as a bounded, tradable object, for instance when encoding indigenous knowledge for the protection of intellectual property; indeed, it signifies a long overdue move to question the assumption of a hierarchy of knowers central to the modernist project. The practitioner's knowledge, however, is situated in immediate experience and direct engagement with everyday tasks, not 'a collection of real entities, located in heads' (Lave, 1993: 12). The attempt to conserve practical knowledge and store it *ex situ* in archives and databases is likely to fail. Not only is it an unrealistic task simply because of the fleeting character and ever-changing nature of the knowledge involved, it will probably reinforce the hierarchy of knowledge which it is trying to avoid for the benefit of textual experts and management elites.

Focusing on practical knowledge represents one attempt to resolve such conceptual ambiguities and theoretical difficulties. Such a focus does not assume a cultural or temporal boundary, the radical separation of producers and scientists, participants and observers, traditionalists and modernists. There may be different ways of knowing, contextualized constructions, on the one hand, and, on the other, de-contextualized abstractions – generalizations across contexts. On some occasions, however, all of us seek to formulate our tacit knowledge in general terms, by verbal or textual means. Likewise, practical knowledge is not restricted to any particular group of people, for none of us (including practising scientists) would manage to live without it.

The theory of practice (see, for instance, Ingold, 1992; Lave, 1988, 1993) offers a view of learning and craftsmanship which is very different from that of orthodox learning theory. Informed by the notions of situated action and mutual enskilment, it emphasizes democratic collaboration and direct engagement with everyday tasks. Such a perspective not only provides a useful antidote to the project of modernist management, it resonates with some aspects of the discourse of Icelandic fishers (Pálsson 1994). For them, 'real' schooling is supposed to take place in actual fishing, not in formal institutions. As one skipper put it: 'Naturally, most of the knowledge one uses on a daily basis is obtained by experience. One learns primarily from the results of personal encounters, that is what stays with you.' Even a novice fisherman, skippers say, with minimal experience of fishing, is likely to know more about the practicalities of fishing than the teachers of the Marine Academy. Therefore, there is little connection between school performance and fishing success. Questioned about the role of formal schooling, skippers often say that what takes place in the classroom (during lessons in astronomy, for instance) is more or less futile as far as fishing skills and differential success are concerned, although they readily admit that schooling has some good points, preventing accidents and promoting proper responses in critical circumstances involving the safety of boat and crew. Such claims seem to be supported by statistical evidence.

A recent study by the present author has found no significant relationship between the catch for a season and skippers' grades in their final examinations in the Marine Academy (Pálsson and Helgason, 1996a). The catch for the season correlated strongly with number of trips and boat size, but not with average grade at school. By this evidence, success at school has no relation to success in fishing. It seems unreasonable to assume that 'fishiness', the ability to catch fish, is given in advance; more likely it is developed later on, in the course of practice in the company of tutors (more experienced skippers) and other crew. There is no relationship between fishing success, on the one hand, and, on the other, skippers' age and years of skipperhood. This suggests, along with ethnographic data, that what matters is the quality of practical experience (in particular, the apprentice–tutor relationship between novice and skipper) rather than its duration.

Skipper education recognizes the importance of practical learning. Earlier participation in fishing, as a deck-hand (*háseti*), is a condition for formal training, built into the teaching programme; this is to ensure minimum knowledge about the practice of fishing. Once students in the Marine Academy have finished their formal studies and received their certificate, they must work temporarily as apprentices – as mates (*stýrimenn*) – guided by a practising skipper, if they are to receive the full licence of skipperhood. The attitude to the mate varies from one skipper to another: as one fisher remarked, 'some skippers regard themselves as teachers trying to advise those who work with them, but others don't'. While skippers differ from one another and there is no formal economic recognition of their role in this respect in terms of a teaching salary, according to many skippers the period of apprenticeship is a critical one. It is precisely in the role of an apprentice at sea that the mate learns to attend to the environment *as a skipper*. Working as a mate under the guidance of an experienced skipper gives the novice the opportunity to develop attentiveness and self-confidence, and to establish skills at fishing and directing boat and crew.

THE 'TRAWLING RALLY'

Recently, attempts have been made at bridging the growing gap between fishers and scientists. An important example is the 'trawling rally' (*togararall*) – a procedure whereby a group of skippers regularly follow the same trawling paths identified by biologists in order to supply detailed ecological information. In an attempt to encourage cooperation between scientists and fishers and to involve the latter in the collection of detailed ecological data on the state of the seas and the fishing stocks, the Marine Research Institute hires a group of skippers to regularly fish along the same pre-given trawling paths on their commercial vessels (see Pálsson, Ó. K. et al., 1989); this is the so-called 'groundfish project' – in everyday language, the 'trawling rally' – initiated in 1985:

> The cooperation with fishermen is based on the main objective of the project; to increase precision and reliability of stock size estimates of relevant fish stocks, especially cod, through the integration of fishermen's knowledge of fish behaviour and migrations, as well as the topography of the fishing grounds.
> (Pálsson, Ó. K. et al., 1989: 54)

Because mature cod sometimes migrate from Greenland to Icelandic fishing grounds, thereby making traditional stock assessment on the basis of fishing statistics relatively unreliable, a systematic fishery-independent survey was seen to be essential. Such a large-scale project was beyond the capabilities of Icelandic research vessels and, therefore, commercial vessels were hired for the task. For two weeks in March every year, five vessels survey the same research stations (595 in the beginning, later on 600), originally selected through a semi-randomly stratified process in cooperation with fishers. The vessels (stern trawlers) are identical in overall equipment and design, in terms of size, fishing gear, engine power, etc. This is seen to be important to ensure comparable data sets, allowing for reliable estimates of changes in the ecosystem from one year to another.

The trawling rally is an interesting experiment. While it is partly a diplomatic endeavour on behalf of the Marine Research Institute in order to reduce the tension that has developed between biologists and fishers in recent years as a result of stringent scientific management, and to improve the image of the Institute among the general public, there is obviously more to the story. No doubt, the trawling rally yields extensive comparative information that the biologists could not possibly gain otherwise, given their limited funding. Nevertheless, as an attempt to cultivate effective interactions between fishers and biologists for management purposes it has significant limitations. To begin with, while the design of controlled surveying has an obvious comparative rationale, it is also a straitjacket, preventing a more flexible and dynamic sensing of ecological interactions in the sea. Skippers fail to be impressed with the scientific design, criticizing the biologists for 'isolating themselves temporarily on particular ships, pretending to practise great science', to quote one of the skippers. Many skippers pointed out during interviews that, fixed to the same paths year after year, the rally fails to respond to fluctuations in the ecosystem, thus providing unreliable estimates; one of them, a skipper who had participated in the trawling rally, remarked that knowing how the biologists worked he had lost all faith in scientific procedures!

Also, the reliance on trawling, skippers say, is likely to produce biased results. Often, those fishing with gill nets on nearby grounds offer a very different picture. From the skippers' point of view, a more intuitive and holistic approach, allowing for different kinds of fishing gear and greater flexibility in time and space, would make more sense. Indeed, skippers discuss their normal fishing strategies in such terms, emphasizing constant experimentation, the role of 'perpetual engagement' (*að vera í stanslausu sambandi*), and the importance of 'hunches' (*stuð*) and tacit knowledge.

Like their colleagues in other parts of the world, Icelandic biologists have often focused on one species at a time. Fishers question their basic assumptions, arguing that understanding of fish
migrations and stock sizes is still very small. Knowledge of the
ecosystem, they claim, is too imperfect for making reliable forecasts. 'Erecting an ivory tower around themselves,' one skipper
argued, 'biologists are somewhat removed from the field of action;
they are too dependent on the book.' While such comments have
to be seen in the light of cultural and economic tension between
social classes and between centre and periphery, they should not
be rejected on that ground alone as they also have some grain of
truth. Biological estimates and fisheries policy are often literal and
rigid in form, unable to deal with variability and to respond to
changes in the ecosystem. Skipper knowledge, in contrast, the result of situated learning, of direct engagement with the aquatic
environment, is necessarily tuned to the flux and momentum of
fishing. An important task on the management agenda should be
to look for ways in which that knowledge can be employed to a
greater extent than at present for the purpose of responsible and
democratic resource management, bridging the modernist gap between scientists and practitioners.

Oddly enough, a recent two-volume survey of the 'history of marine
research' on Icelandic fishing grounds by Jónsson (1988, 1990), former
director of the Marine Research Institute, only has one page on
the trawling rally (1990: 122–3). The negligible attention paid to
the trawling rally in Jónsson's extensive analysis should not be surprising, however, given the discursive tone and framework of his
general account, focusing on the triumph of modern science over
the ignorance of folk understanding. Jónsson mentions that research
stations were defined 'equally' by skippers and marine biologists
and that the former were guided by 'experience', but his comment
on the reasons for resorting to skippers' experience is surprisingly
blunt and restricted, emphasizing the sheer amount of data required
and the practical constraints of their collection – the fact that 'the
Marine Research Institute did not have at its disposal the fleet
required for such a data collection' (Jónsson 1990: 122). Despite
the occasional lip service in the reference to 'collaboration' (*samráð*),
there is little real dialogue between fishers and marine biologists.

CONCLUSION

Discourse is necessarily embedded in social relations. It is precisely through social discourse that people redefine their relations to one another and their place in the world; discourse and context are dialectically interlinked. The significant shift in the way in which Icelanders talk about knowledge of the marine ecosystem and the fisheries, therefore, alerts us to fundamental changes in social relations. Science is no longer a humble, somewhat naive, project, an extension of outdoor learning, but the imposition of a particular world-view. Not only are some of the participants in the current regime (environmental 'experts') presented as the managers or stewards of marine resources, positioning themselves as objective spectators outside the realm of predictable nature, but parts of the ecosystem are defined as commodities, parcelled out to individual producers and, finally, marketed among independent transactors.

This development is not unique to Iceland. The regime of the aquarium is part and parcel of the modernist regime. Aquariums are human constructs, designed and managed for human purposes. Moreover, aquariums usually owe their construction to the fascination with single species and individual animals. They may be difficult to administer at times and the results are not necessarily along the lines envisaged by those in charge, but it is generally assumed that things are 'under control'. Like keepers of aquariums, marine biologists have typically focused on one species at a time – modelling recruitment, growth rates and stock sizes – although recently they have paid increasing attention to analyses of interactions in multi-species fisheries.[2] Fisheries management seeks to systematically affect the structure of fish populations by controlling the relative sizes of different species and year classes, through regulations concerning fishing effort, gear, mesh sizes, territorial restrictions, etc. In practice, marine scientists habitually present ecosystems as predictable, domesticated domains, despite the uncertainties involved (Finlayson, 1994). They may qualify their analyses and predictions with reference to the 'margin of error', but pressured by politicians and policy-makers they tend to act as if that margin is immaterial. Such a modernist approach to resource management draws upon textual notions of scientific practice which developed during the Middle Ages when it was customary to speak of nature as 'God's book' and to regard science as the 'reading of the book of nature'.

Not only is nature presented as an inherently textual and logical domain, the project of the resource economist and manager is sometimes likened to that of the engineer or the technician. Somewhat surprisingly, with the modernist idea of the engineering of the oceans and resource management there was a rather sudden shift in Western attitudes to aquatic mammals (Kalland, 1993): whales ceased to be a resource – just-like-any-other-fish – and became quasi-human beings, sometimes 'adopted' by environmentalists. Adopted whales need not, however, be antithetical to the modernist regime. Giant marine mammals seem to be ideally suited for taking on the role of the goldfish in the great aquarium of the ocean.

It may be tempting for social scientists to either submit to the populist notion that textualizes and privileges the traditional and indigenous or to contribute to the opposite enterprise, the reproduction of the master narrative of science. A more realistic and democratic approach, however, would be to search for an egalitarian discursive framework akin to the ethics of the 'ideal speech situation' identified by Habermas (1990: 85), a communicative strategy for recognizing differences and solving conflicts in the absence of repression and inequality. While the trawling rally represents an interesting endeavour at bridging the gap between fishermen and biologists, it has its shortcomings. It is important to look for alternative ways of engaging fishers, of using knowledge obtained in the course of production for the purpose of responsible resource use.

In recent years, the dualist theory of knowledge on which the modernist production regime is based has been challenged on a number of fronts. For one thing, practical knowledge has been firmly placed on both the theoretical and management agenda. The current tendency, however, to reify practical knowledge as a marketable commodity has certain drawbacks: in particular, it fails to address the situatedness of the practitioner's knowledge. At the same time, the notion of the absolute objectivity of science, the idea of some scientific Archimedean standpoint outside nature and history, is frequently subject to critical discussion, with the growing awareness that the modernist perspective fails to account for the actual practice of modern science. As Latour (1994) argues, modern science has never been able to meet its own criteria; paradigms and *épistémès* are inevitably social constructs, the products of a particular time and place. Scientific knowledge, of course, involves practical knowledge obtained in the course of engagement and experimentation. The idea of the aquarium, then, may not be so misleading after all. If it is to make some sense, however, the aquarium will have to include the practitioners of science (Pálsson, 1996). In such a revised sense, the metaphor of the aquarium would come close to the Greek idea of *oikos*, the 'house'.

NOTES

1. The study on which the article is based relates to two collaborative research projects, 'Common Property and Environmental Policy in Comparative Perspective', initiated by the Nordic Environmental Research Programme, and the 'Property Rights Program' of the Beijer Institute of the Swedish Academy of Sciences, coordinated by Susan Hanna. It has also received financial support from the Nordic Committee for Social Science Research (NOS-S), the Research Centre of the Vestman Islands and the University of Iceland, and the Icelandic Science Foundation. I thank Jónas G. Allansson (University of Iceland) and Agnar S. Helgason (University of Oxford) for their help with interviews and data collection.
2. The focus on single animals and single species is not restricted to Western, scientific accounts. Carrier (1987) describes the case of the Ponam in Papua New Guinea, emphasizing the differences between Ponam accounts of single species and modern ecological models. In other respects, however, the Ponam do not seem to share the scientific view of the 'aquarium'; in particular, they do not seem to be preoccupied with human control and predictability.

BIBLIOGRAPHY

Ægir, 1905–95 (Reykjavik) (Journal of the Icelandic Fisheries Association).
Agrawal, A. (1995) 'Dismantling The Divide Between Indigenous and Scientific Knowledge', *Development and Change*, 26: 413–39.
Árnason, R. (1993) 'The Icelandic Individual Transferable Quota System: A Descriptive Account', *Marine Resource Economics*, 8: 201–18.
Brush, S. (1993) 'Indigenous Knowledge of Biological Resources as Intellectual Property Rights: The Role of Anthropology', *American Anthropologist*, 95(3): 653–86.
Carrier, J. (1987) 'Marine Tenure and Conservation in Papua New Guinea: Problems in Interpretation', in McCay, B. J. and Acheson, J. M. (eds), *The Question of the Commons: The Culture and Ecology of Communal Resources* (Tucson: University of Arizona Press), pp. 142–70.
Chapin, M. (1994) 'Recapturing the Old Ways: Traditional Knowledge and Western Science among the Kuna Indians of Panama', in Kleymeyer,

C. D. (ed.), *Cultural Expression and Grassroots Development: Cases from Latin America* (Boulder, Colo. and London: Lynne Rienner), pp. 83–101.

DeWalt, B. R. (1994) 'Using Indigenous Knowledge to Improve Agriculture and Natural Resource Management', *Human Organization*, 53(2): 123–31.

Finlayson, C. (1994) *Fishing for Truth* (St John's, Newfoundland: ISER).

Foucault, M. (1973) *The Birth of the Clinic: An Archaeology of Medical Perception*, trans. Sheridan, A. M. (London: Tavistock).

Gudeman, S. (1992) 'Remodelling the House of Economics: Culture and Innovation', *American Ethnologist*, 19(1): 141–54.

Gudeman, S. and Rivera, A. (1990). *Conversations in Colombia: The Domestic Economy in Life and Text* (Cambridge: Cambridge University Press).

Habermas, J. (1990) 'Discourse Ethics: Notes on a Program of Philosophical Justification', in Benhabib, S. and Dallmar, F. (eds), *The Communicative Ethics Controversy* (Cambridge, Mass.: MIT Press), pp. 60–110.

Hutchins, E. (1995) *Cognition in the Wild* (Cambridge, Mass.: MIT Press).

Ingold, T. (1992) 'Culture and the Perception of the Environment', in Croll, E. and Parkin, D. (eds), *Bush Base: Forest Farm: Culture, Environment and Development* (London: Routledge) pp. 39–56.

Jónsson, J. (1988) *Hafrannsóknir við Ísland I: Frá öndverðu til 1937* (Reykjavik: Bókaútgáfa menningarsjóðs).

Jónsson, J. (1990) *Hafrannsóknir við Ísland II: Eftir 1937* (Reykjavik: Bókaútgáfa menningarsjóðs).

Kalland, A. (1993) 'Whale Politics and Green Legitimacy: A Critique of the Anti-Whaling Campaign', *Anthropology Today*, 6, December, pp. 3–7.

Latour, B. (1994) *We Have Never Been Modern* (Cambridge, Mass.: Harvard University Press).

Lave, J. (1988) *Cognition in Practice: Mind, Mathematics and Culture in Everyday Life* (Cambridge: Cambridge University Press).

Lave, J. (1993) 'The Practice of Learning', in Chaiklin, S. and Lave, J. (eds), *Understanding Practice: Perspectives on Activity and Context* (Cambridge: Cambridge University Press), pp. 3–32.

McGoodwin, J. R. (1990) *Crisis in the World's Fisheries: People, Problems, and Policies* (Stanford, Calif.: Stanford University Press).

Macinko, S. (1993) 'Public or Private? United States Commercial Fisheries Management and the Public Trust Doctrine, Reciprocal Challenges', *Natural Resources Journal*, 32: 919–55.

Marx, K. (1976)[1867] *Capital, Vol. 1*, trans. Fowkes, B. (Middlesex: Penguin).

Pálsson, G. (1991) *Coastal Economies, Cultural Accounts: Human Ecology and Icelandic Discourse* (Manchester: Manchester University Press).

Pálsson, G. (1994) 'Enskilment at Sea', *Man*, 2, 9, 4: 901–28.

Pálsson, G. (1996) 'Human–Environmental Relations: Orientalism, Paternalism, and Communalism', in Descola, P. and Pálsson, G. (eds), *Nature and Society: Anthropological Perspectives* (London and New York: Routledge).

Pálsson, G. (1997) Learning by Fishing: Practical Engagement and environmental concerns', in Berkes, F. and Folke, C. (eds), *Linking Ecological and Social Systems* (Cambridge: Cambridge University Press).

Pálsson, G. and Helgason, A. (1995) 'Figuring Fish and Measuring Men:

The Quota System in the Icelandic Cod Fishery', *Ocean and Coastal Management*, 2, 8, 1–3: 117–46.

Pálsson, G. and Helgason, A. (1996a) 'Schooling and Skipperhood: Bodies of Knowledge and Communities of Practice' (unpublished paper).

Pálsson, G. and Helgason, A. (1996b) 'The Politics of Production: Enclosure, Equity and Efficiency', in Pálsson, G. and Durrenberger, E. P. (eds), *Images of Contemporary Iceland: Local Lives and Global Contexts* (Iowa City: University of Iowa Press).

Pálsson, Ó. K. et al. (1989) 'Icelandic Groundfish Survey Data Used to Improve Precision in Stock Assessments', Journal of Fisheries 9: 53–72.

Schopka, S. (1975) 'Fiskispár' ('Fishing forecast'), *Vikingur*, pp. 47–9.

Scott, A. D. (1989) 'Conceptual Origins of Rights Based Fishing', in Neher, P. A., Arnason, R. and Mollet, N. (eds), *Rights Based Fishing* (Dordrecht: Kluwer Academic), pp. 11–45.

Wilson, J. A. et al. (1994) 'Chaos, Complexity and Community Management of Fisheries', *Marine Policy*, 18(4): 291–305.

15 Fishing and Fairness: the Justice of the Common Fisheries Policy[1]

Tim S. Gray

INTRODUCTION

This chapter discusses the normative rather than the empirical dimensions of the Common Fisheries Policy (CFP). It focuses on the ethical or moral issues raised by questions of equity or distributive justice. The reason why I have chosen this topic is partly because little or nothing has been published on it, and partly because fairness is a very important element of the CFP, highly relevant to its acceptability to fishers.[2] It is often argued that the fairness of the CFP is a precondition of its compliance, but rarely, if ever, is that fairness examined. The aim of this discussion is to help fill this gap.

There are at least four different criteria by which the CFP could be judged: efficiency of production of fish (market criterion); maximization of employment (labour criterion); protection of fishing communities (social criterion); maintaining fish stocks (conservation criterion). But underlying all of these is a fifth criterion: establishing a morally fair regime (ethical criterion). This oft quoted (e.g. Symes and Crean, 1995: 395–8; Wise, 1984: 142, 149–51; CFPRG, 1996 Press Release: 4), but rarely analysed, criterion is the one I want to explore by asking the following questions. What are the principles of fairness which drive the CFP? Are they coherent in themselves? Do they form a consistent set of moral priorities?

To help us in evaluating the principles of fairness underlying the CFP, it is necessary to consider the two ways in which principles of fairness are conceived in political philosophy: the procedural and the substantive. Procedural principles of fairness are rules which lay down prescribed ways of dealing with claims, the assumption being that provided the given rules are adhered to, a fair outcome can be guaranteed. Procedural fairness embraces, for example, legal process (whatever the law's course requires), natural justice (no discrimination or bias) and consent (everyone agrees to the rules).

However, critics may argue that even if the correct procedures have been followed, the outcome may not be fair as judged by substantive principles of fairness. At any rate, in the case of fishing, unlike, say, a lottery, procedural regularity is not all there is to issues of fairness. The rules must pass the test of substantive, as well as procedural, fairness. Substantive principles of fairness are concepts that are chosen because they embody fairness within them. Examples of substantive fairness include: equality (everyone given the same); need (allocations varying with different welfare needs); entitlement (people obtain what they are historically entitled to); desert (individuals get what they deserve).

What I want to argue is that we can see elements of virtually all of these procedural and substantive principles in one form or another in the CFP, and that the only way in which we can judge between them is to weigh up their respective strengths and weaknesses comparatively in the contexts in which they are deployed and in the light of our own value systems. There are two main principles that drive the CFP: the principle of equal access and the principle of relative stability. Let us consider them in turn, and examine their claims to procedural and substantive fairness.

THE PRINCIPLE OF EQUAL ACCESS

This is the primary principle of the CFP: the basic rule on which it is founded. What does the principle of equal access mean? It is a rule which arises out of the Treaty of Rome's principle of non-discrimination among member states, adopted in 1970 at the inception of the CFP. It is set out as follows:

> Rules applied by each Member State in respect of fishing in the maritime waters coming under its sovereignty or within its jurisdiction shall not lead to differences in treatment of other Member States. Member States shall ensure in particular equal conditions of access to and use of the fishing grounds situated in the waters referred to ... for all fishing vessels flying the flag of a Member State and registered in Community Territory.
>
> (CFPRG, 1996, II: 3)

What the principle of equal access means is equality of access for member states to Community waters, that is access on equal terms for all member states – not freedom of access, that is unconditional access for all member states. Member states retain the right to impose restrictions on access to their waters, 'providing that these measures applied equally to all Community fishermen regardless of nationality' (Wise, 1984: 91).

HOW FAIR IS THE PRINCIPLE OF EQUAL ACCESS?

Procedural fairness

The manner of the adoption of the principle of equal access has been strongly criticized. According to Holden, it was rushed through by the original six members of the EC immediately before accession talks began with Denmark, Finland, Norway and the UK in 1970.

All four applicants had major fisheries interests and much of the fishing by three of the original six Member States took place in their waters. If the 'six' did not agree a policy before the applicants joined, the new Member States would be able to influence its development, in particular the conditions of fishing and access to national waters. With a policy in place prior to their accession, the new Member States would have to accept the *acquis communitaire* [*sic*]. The 'six' just managed to agree, almost literally at the eleventh hour.   (Holden, 1994: 19)

Such a process seems to breach at least one of the rules of procedural fairness – that of natural justice – in that it was clearly designed to discriminate in favour of existing member states at the expense of the new states.

However, it could be argued that since the 'new' states had not yet joined the Community, such last minute 'rigging' of the rules in the interests of existing member states was not unjust, since it is entirely fair that the rules of a club should be determined by its existing members. The six were fully entitled to make whatever rules suited them, and so there was nothing procedurally improper about their decision. After all, the four new applicants were not obliged to join the EC; they could have declined to join on grounds that they rejected the 'rigged' rules (indeed, Norway did just that). In joining, Denmark, Ireland and the UK consented to all EC treaties and rules – including the principle of equal access. Hence, it could be argued that the principle could not be procedurally unjust: prior consent is a guarantor of procedural fairness.

On the other hand, it could be argued that since the six knew that the four were about to join, the way they used the procedure may have been fair in letter, but not in spirit. Moreover, it has been claimed that while it may be true that the UK government consented to the equal access principle in 1970, UK fishers did not. Indeed, Eurosceptics claim that the UK electorate did not consent: they were not even consulted by the then Prime Minister, Edward Heath. But such an argument opens up complex issues of representative versus direct democracy; suffice it to say that in a representative system (such as in the UK), on almost any question, a government's policy could be judged to lack majority consent, let alone unanimous consent, but not to lack electoral legitimacy.

Moreover, the equal access principle may be deemed to pass the test of procedural fairness in that, at least from 1983 to 1992, it was unchallenged by Fisheries Ministers: 'Access has never once been the subject of debate in the Council. . . . The provisions on access were virtually unquestioned for ten years, from 1983 to 1992. The Council adopted them unchanged to continue another ten years, until 31 December 2002' (Holden, 1994: 117, 118). However, it must be noted that the question of access is not the sort of issue that one would expect to be raised in the Council of Fisheries Ministers; debate on such a fundamental issue would be part of the agenda for an Intergovernmental Conference.

Substantive fairness

The equal access principle may satisfy tests of procedural fairness (it may have been introduced in a fair way) but it may nevertheless fail the test of substantive fairness (it may not be fair in itself). The substantive concept of fairness to be found in the equal access principle is, of course, the concept of equality. All member states have equality of access to each other's waters, that is, since 1977, within their 200-mile exclusive economic zones (EEZs). But on what basis is the concept of equality justified here? It may be justified on three grounds:

(a) The EU depends on reciprocity – if fishers of one member state want access to waters of another member state, they must allow reciprocal access by foreign fishers to their own waters.

(b) The waters of the EU are a Community resource – a common pool or pond – and as such ought to be open to all on equal terms.

(c) The EU is founded on the principle of a level playing field for all nationals of member states.

However, all three justifications are controversial. Taking (a) first, according to Wise, there are three forms of reciprocity entailed in the equal access principle. First, there is reciprocity between fishers. For example, if UK fishers want to fish within the waters of France, as a quid pro quo they must accept that it is only fair that French fishers should be allowed to fish in British waters. The moral bankruptcy of a chauvinistic position which rejects such reciprocity is exposed by Symes and Crean: 'The fishermen tend to face in two directions: they oppose any curtailment of their rights of equal access to stocks in European waters and, at the same time, resist the invasion of their perceived national territories by foreign fishermen in pursuit of identical rights of access' (Symes and Crean, 1995: 398). However, an objection to this argument is that the reciprocity here is very one-sided: since British waters contain fish stocks which are approximately five times as abundant as the fish stocks in French waters, British fishers are 'giving' to the French much more than they are 'getting back' from the French. This does not seem fair.

Perhaps the second form of reciprocity redresses this imbalance? This is reciprocity between access to fishing areas and access to fishing markets.

> Was it not basic justice, argued its protagonists, that the free movement of fish products within the EEC be counterbalanced by equal conditions of entry to fishing grounds? Why should Britain and Ireland deny continental fishermen access to waters they had traditionally exploited and then expect to export freely to other member states where there would be an increased demand for imports following decreased landings from domestic fleets cut off from fisheries around the British Isles? (Wise, 1984: 165)

On this argument, we cannot expect to be treated with goodwill by foreign consumers if we are perceived to treat their nationals discriminately. While this form of reciprocity is not one-sided (the extent of the access to foreign markets will depend on the extent of the domestic product), it is a contingent rather than an ethical argument. That is to say, it amounts to saying not that we have a moral duty to allow foreign nationals access to our waters if we want to sell them our fish, but that we would be acting prudently to do so in order to avoid a foreign consumer backlash.

The third form of reciprocity to which Wise draws attention may be summed up in the familiar injunction 'to refrain from doing to others what you object to others doing to you'. For example, 'Why . . . should continental fishermen be evicted from their traditional fishing grounds to make space for British distant-water vessels expelled from theirs around Iceland and elsewhere?' (Wise, 1984: 166). The point being made here is that it is unfair of one member state to respond to discrimination against its nationals by a third state by itself discriminating against the nationals of a fellow member state. Two wrongs do not make a right. To suffer injustice is not a justification for inflicting the same injustice.

This third form of reciprocity seems to me to be the strongest form. It can be broadened into the Rawlsian idea of asking people, 'how would you feel if you were in the other person's shoes?' It appeals to fishermen to take a detached, non-chauvinistic view of the issue. From the Rawlsian original position (Rawls, 1972), the equal liberty principle would be chosen in order to safeguard oneself from discrimination. This is a powerful justification for equality of access.

Turning now to (b) the second justification for the equal access principle – that EU waters are a Community resource – we must note that the notion of a common pool of the waters of all the EU member states is anathema to many critics who argue that there is no other resource in the EU over which the idea of a common resource is superimposed. The critics point out, for example, that the Common Agricultural Policy (CAP) does not assume that agriculture in member states' land is a common pool resource of the EU, so why should there be one in the waters of member states? The answer to this objection is twofold, according to the CFP's defenders. First, fish do not confine themselves to national boundaries, and this justifies their designation as a common, rather than a national, resource. Second, there are other resources in the EU that could be described as common pool – such as air and wild birds – 'access' to both of which the EU regulates on an equal basis. However, even if EU waters and the fish within them are legitimately designated as a common resource, the question remains open of whether or not equal access to that common pool resource is fair.

This brings us to (c) the third ground for equality of access – that of a level playing field for all EU nationals. This notion is of course enshrined at the heart of the EU, incorporated in the idea of a single market. However, critics have argued that there is not at present a level playing field in EU waters, in that some member states' fleets are stronger than others, and that the weaker member states' fleets require special protection to enable them to compete on more equal terms. This is certainly the argument put forward to justify restrictions on foreign vessels in the Irish Box, on grounds that Ireland's infant fisheries industry required protection. It may also be the justification for the 1972 derogation from equal access in respect of foreign vessels in the coastal areas (six- and 12-mile limits) of all member states (to protect small inshore vessels against large, distant-water vessels).

In other words, most of the exceptions (derogations) from the equal access principle could be justified on grounds that they were necessary to create conditions of genuine equality between member states' fleets. On this argument, the egalitarianism of the equal access principle is preserved, not diminished, by the derogations. If, however, such an argument appears too paradoxical to be convincing (inequality guarantees equality), an alternative way of interpreting the derogations is to say that the inequalities were justifiable – that is, that the egalitarian content of the equal access principle has been diluted for good reason. One good reason for diluting the equal access principle might be the greater degree of dependence on fishing of the above areas. However, as Wise points out: 'Continental states ... resisted the argument that the "dependence" of certain British and Irish regions on fishing justified an abandonment of the national non-discrimination principles underpinning the CFP. They made the obvious, but often missed, point that all fishermen are "dependent" on fishing' (Wise, 1984: 167). All fisheries have a vital interest in fishing: how do we judge which interests are most important? (This issue will be explored later in relation to the 'vital needs' element of the principle of relative stability).

Another good reason for diluting the equal access principle might be the greater proximity of the Irish and British fleets to these areas. The concept of proximity entails that closeness to a resource generates a preferential right to exploit that resource. On this view, because Ireland and the UK are closer than other member states to the fish stocks in the coastal areas and EEZs around the British Isles and Ireland, they have a prior right to exploit those stocks. This right entitles them to special treatment of the kind that we have seen they have obtained.

However, the concept of proximity as a substantive principle of fairness is highly controversial. Why should mere closeness to a resource generate a moral claim to preferential access in exploiting it? 'Closeness' in itself is simply a random physical fact of no moral justification in itself. Being born, or living, near some resource is a purely accidental or contingent matter; how can such a physical fact generate a moral claim? This is to commit the so-called 'naturalistic fallacy' – attempting to deduce an 'ought' from an 'is'. Surely, the critics argue, it is only if and when someone does something to the resource that they generate a moral claim to it. Hence, for example, it is only when UK fishers actually fish the stocks in the UK waters that they can in any sense be held to create a proprietorial right to those fish stocks. But, of course, by the same argument, when French fishers fish the same stocks in UK waters, they too must be held to create a proprietorial right to those fish stocks. In reconciling these two sets of rights, some principle such as relative stability would have to be devised – that is, a principle which would adjudicate between them not on grounds of proximity, but on grounds of differential use, for example by historic catch records.

However, it could be argued that while proximity in itself is morally irrelevant, it may be less so if we regard it as an extension of national property rights. If we accept that each country has ownership rights over the resources within its territory, the question becomes 'what is a nation's territory?' If its territory is deemed to include the seas up to 200 miles around its coastline, it follows that it owns those seas and their contents. However, while such an argument might be persuasive in relation to inert products such as oil or natural gas or minerals on the sea bed, it is less persuasive in the case of fish since, as noted earlier, fish move in and out of 'territorial' waters, spawning in one area and migrating to others. No single state could fairly claim exclusive ownership rights over migratory fish – they 'belong' to several states, or perhaps to none. So the argument of proximity is no more convincing than the argument of dependence in justifying derogations from the equal access principle.

Conclusion on the equal access principle

Clearly, there are competing notions of what the equal access principle entails in terms of fairness, both procedurally and substantively. The way in which it was introduced seems unfair in spirit if not in letter, and there are some powerful reasons for considering granting limited exceptions to its application. Our verdict on its fairness must, therefore, be conditional – that it is fair procedurally and substantively only up to a certain point. Perhaps fairness is secured when it is combined with the principle of relative stability? This is the issue to which we must now turn.

THE PRINCIPLE OF RELATIVE STABILITY

This is the secondary principle of the CFP, the main restraint on the primary principle of equal access. What does the principle of relative stability mean? It is a rule adopted in 1983 for dealing with the problem of scarcity of the fish stocks. It operates by distributing shares of total allowable catches (TACs) largely on the basis of historic patterns of catches, adjusted for need and compensation.

> In 1983, Council Regulation 170/83 . . . introduced the concept of relative stability (or fixed percentage shares) in quota allocation, within a TAC framework and on the basis of historic fishing patterns (modified as necessary by arrangements to protect fisheries dependent communities (Hague Preference) and compensation for lost international opportunities).
>
> (CFPRG, 1996, II: 4)

It differs from the equal access principle in that it was designed to regulate takes of fish stocks, not conditions of access to fishing areas; under the relative stability principle, the good to be distributed is the total fish catch, not the opportunity to engage in fishing. However, since the principle of relative stability regulates takes of fish stocks within each area, it effectively does control access to many areas, and in this sense it represents both a departure from, and a qualification of, the equal access principle (Deas, 1996).

The principle of relative stability was introduced, first, to avoid annual wrangling over the allocation of TACs, and second to provide a degree of security for all member states' fishing industries by assuring them a guaranteed share of fish. Since, however, it was impossible to guarantee absolute quantities of fish, because of the wide fluctuations in the biomass from year to year, the Commission 'proposed that each member state should be guaranteed only a specified percentage of the TAC for each stock, to which the term "relative stability" was given' (Holden, 1994: 43). These percentages were calculated on the basis of three factors: 'historical catches', 'vital needs' ('Hague Preferences') and 'compensation for jurisdictional losses'. The main factor was the first, historical catches, and referred to the track record of catches during the period 1973–8: 'The year 1978 was selected as the end of the period because it was, at the time, the last year for which official catch statistics were available; the Commission was determined to use figures which were published [by ICES] . . . and which could not therefore be disputed' (Holden, 1994: 43).

The second factor – 'vital needs' or 'Hague Preferences' (so named because they were set out at a 1976 Council meeting in The Hague) – referred to special allowances given to areas flanked by coastal regions which relied heavily on fishing. Regulation No. 170/83 states that relative stability 'must safeguard the particular needs of regions where local populations are especially dependent on fisheries and related industries' (Holden, 1994: 118). These areas included the whole of Eire (which had its 1975 catch quantities doubled) and the whole of Scotland, Northern Ireland and north-eastern parts of England as far south as Bridlington (which had their 1975 catches by vessels up to 24 metres long doubled). The third factor – 'compensation for jurisdictional losses' – refers to losses suffered by member states' fishing fleets by non-member states extending their fisheries' limits to 200 miles in 1977. This compensation took the form of extra fish quotas in EU waters.

HOW FAIR IS THE PRINCIPLE OF RELATIVE STABILITY?

Procedural fairness

The first procedural point to be noted about the relative stability principle is its perceived importance for the CFP. According to the CFP Review Group, the principle of relative stability is much more important than the principle of equal access. 'Access to waters is largely meaningless without access to fish quotas. Relative stability is thus a vitally important and practical concept, whereas access to waters is of less practical significance' (CFPRG, 1996, II: 9). Indeed, Holden claims that 'the principle of relative stability . . . is now considered the fundamental, untouchable keystone of the conservation policy' (Holden, 1994: 256). This reflects the widespread popularity of the principle: 'The [1992] review showed that the principle of relative stability has, for the majority of Member States, proved politically totally satisfactory. Only one Member State is wholly discontent. Spain voted against Regulation No. 3760/92' (Holden, 1994: 122).

This suggests that the principle of relative stability largely meets one of the tests of procedural justice – namely that it is founded on the consent of all (but one) of the parties affected. However, this suggestion is disputed in some quarters. For instance, Spain and Portugal argue that they should be given greater quotas for fish stocks, to take account of two factors: first, that they had a long historical track record of large catches in, for example, Greenlandic waters, before 1977, for which they claim compensation under the CFP scheme for jurisdictional losses; second, that other member states are not catching their full quotas in those areas and that the shortfall should be reallocated to Spain and Portugal. However, both these claims were rejected by the European Court of Justice. First, under Article 2 of the Treaty of Accession, Spain and Portugal had accepted the *acquis communautaire*, which included the principle of relative stability as defined in Regulation No. 170/83, by which no account was taken of Spanish and Portuguese jurisdictional losses. Second, whether or not member states catch their full quotas is irrelevant to the allocation of percentages. In any case, it could be argued that even if two countries are unsatisfied by some aspects of the relative stability principle, the vast majority of member states are satisfied, and majority consent (not unanimity) is all that is required for procedural fairness.

However, each of these issues raises important questions of procedural injustice (and arguably also of substantive justice) in that the objections lodged by Spain and Portugal are aimed not only at the procedure whereby the above decisions were reached, but also at the content of those decisions. First, is it fair that new member states cannot claim any compensation for their jurisdictional losses resulting from the 200-mile EEZs adopted by the United Nations Convention on the Law of the Sea (UNCLOS) in 1977? Doesn't this mean that new member states are treated differently from 'old' member states for no good reason other than the self-interest of the old member states? Of course, it is true that Spain and Portugal knew these terms before they signed their Treaties of Accession, therefore there is no breach of one criterion of procedural fairness – consent. However, there does seem to be a breach of another criterion of procedural fairness – that of natural justice – treating like cases in a like manner.

In the eyes of the existing member states, of course, such a procedure is perfectly fair, since they feel they should not be disadvantaged in any way by the accession of new member states. As the CFPRG put it:

> ... UK interests have not been compromised by the accession of new Member States, since the principle of relative stability has simply been applied to the fisheries resources that they bring with them into the Community. There has been no question of parcelling out existing resources to new Members ... relative stability will continue to ensure that the national quota shares of all Member States (including the UK) are protected in the event of any enlargement of the EU. For example, no Member State has suffered from the accession of Finland and Sweden.
>
> (CFPRG, 1996, II: 10, 49)

However, such discrimination against new member states, even if perfectly legal and tacitly consented to by the new states, seems morally dubious. Moreover, as Holden suggests, resentment by new members against this procedural injustice could eventually destroy the principle of relative stability:

> The accession of new Member States to the Community will inevitably create pressure to abandon the principle of relative stability. These new Member States, no more than Spain and Portugal, will not be prepared to accept a system which benefits the 11 Member States which adopted the original conservation policy and disadvantages all others. Sooner or later, the principle will be successfully legally challenged, even though Spain and Portugal have been unsuccessful to date. (Holden, 1994: 241)

The second issue of procedural justice raises the question of the validity of a principle which ossifies allocations on the basis of a historic period of catches recorded in 1973–8. Why should that five-year period be taken for all time as the source of a fair allocation? There are both procedural and substantive questions involved here, but the central issue is that the allocative criterion is arbitrary, not principled, in its application. For one thing, some of the records taken during that five-year period may have been aberrant. For instance, the CFP Review Group drew attention to a UK grievance over the allocation of saithe. 'On North Sea and West of Scotland saithe . . . there are complaints that the UK share, which was based on a very poor historical fishing performance during the late 1970s, is now far too low' (CFPRG, 1996, II: 10). It seems unfair to take one time-slice as the permanent allocative guide in the face of the fluidity of fishing patterns. What ethical relevance to quotas to be allocated in 1996 do catches have which were recorded 20 years earlier? For another thing, if a member state is continuously catching less and less than its quota, why should its percentage not be reallocated to other member states? On this argument, there should be some procedural mechanism whereby adjustments can be made to reallocate unused quotas.

However, there are two arguments for leaving the quota allocation system as it is. First, readjustment could reopen the entire debate about the allocation keys. The CFP Review Group argued that reopening that debate would not benefit Britain.

Each Member State is likely to be dissatisfied on some points of details; but overall the current relative stability arrangements are probably more beneficial to the UK than anything that could now be renegotiated within the CFP . . . The original basis of historic performance was an equitable starting point; and a new debate from first principles would simply invite others to press for improvements at the UK's expense.

(CFPRG, 1996, II: 11)

But, of course, this is a political argument, not an argument from procedural justice. Second, there may be a good reason why a member state fails to catch its full quota (e.g. poor weather conditions, lower biomass of fish, effort restructuring) and it would not be procedurally safe to permanently reduce its quota for factors which may be of temporary duration.

Interestingly, the CFPRG suggests an alternative way of addressing this problem – by bilateral deals with member states who consistently under-fish their quotas (such as France in its North Sea saithe quota (CFPRG, 1996, II: 11)). On procedural grounds, there could be no objection to such bilateral deals, in so far as they were based on mutual consent. However, it raises the question of whether all other member states should be given an opportunity to participate in such deals.

The final procedural issue raised by the principle of relative stability relates to 'quota hopping'. 'Quota hopping' is the term applied to owners in one member state who buy vessels in another member state and, after obtaining fishing licences, can use these foreign-owned flagged vessels to fish against the quota of the home member state. Quota hopping is a very serious problem in the UK, where over a quarter of the tonnage is now foreign-owned, because much of the economic benefit of these flagships flows abroad: many vessels do not land fish in the UK, do not use British shipbuilding or repair yards, and do not employ British fishermen.

However, there can be no objection to quota hopping on procedural grounds. Ownership of foreign vessels is perfectly legitimate within the EU rules – indeed, it is unequivocally sanctioned by the right of establishment. Moreover, it satisfies the rationale of procedural fairness in the processes of market transactions; if people start with legitimate property rights and engage in voluntary exchanges of their assets, then the result of those exchanges must be deemed to be just because it is the outcome of starting points and processes that are themselves just. Whether it violates any substantive principle of justice is, however, another matter, which we address below.

Substantive fairness

The principle of relative stability embodies within it two distinct concepts of substantive justice: entitlement (including compensation) and need.

Entitlement
The primary concept is entitlement. This concept is at the heart of the principle of relative stability, entailed in the very notion of historic catch records. Fishers who have built up a track record of fishing during 1973–8 in a particular area for a particular stock of fish have created an entitlement or prima facie right to continue

to do so, or to receive compensation if they are prevented from doing so. This historic catch concept of entitlement is also the criterion driving the UK's distribution of the quota allocations to producer organizations (POs) from the TACs awarded to the UK by the EU Council of Ministers, though the reference period here is the rolling previous four years.

It is worth pausing to consider the nature of this entitlement argument. It is based on the idea that expending labour not only entitles a person to the fruits of that labour, but also confers a preferential right to continue to expend that labour and reap its fruits in the future. But what is the moral justification of such a right? If the right is granted, it means that other fishers who wish to expend their labour in the same way are disadvantaged, but is this fair? Why should preferential treatment be given to a fisher who has built up a track record? Is it because he was there first, and can claim the right of first 'occupancy' on grounds of 'first come, first served' or 'finders, keepers' or 'possession is 9/10ths of the law'? None of these popular aphorisms seems morally persuasive, since the fact that a fisher was first on the scene was a contingent matter of chance or accident and not a matter of ethical significance. Perhaps the justification lies in the fact that since the fisher has regularly fished in the area, he has 'used' it, invested his labour in it, mixed his labour with it (as John Locke (1962) put it) and therefore, in some sense 'owns' it. But as Robert Nozick (1974) has said, why should mixing one's labour with a resource generate a right to ownership of that resource? Surely it might merely mean that one has wasted one's labour? In any case, Locke's argument was that, since through one's labour one made the land [more] productive, one was entitled to property ownership of the land one had worked on. But in the case of fishing, no such claim could be made – indeed fishers arguably make the sea less productive. As far as fishing is concerned, the Locke argument applies only to aquaculture, where fishers are responsible for 'growing' the fish.

Perhaps the answer is that expending one's labour on something means that one deserves reward for that effort? In other words, labour effort signifies desert. Desert is a strong principle of justice or fairness, but it is also a very difficult concept to work with. It is easy to see how the desert concept can justify ownership of the fish that a fisher has caught: he deserves to be able to sell the catch, it is a just (that is, well deserved) reward for his efforts. But it is not easy to see how desert can justify preferential rights over future fishing stock. Desert is essentially a backward-looking concept: reward for some past effort. Why should that reward include a right to exclude others from making similar efforts in the future? Perhaps the answer is simply that of tradition – a traditional or established pattern of use is one that doesn't need justifying: it is self-justifying. Simply because fishers have traditionally fished in an area entitles them to preferential rights to continue to do so. This is the argument of 'prescriptive rights' advanced by Edmund Burke (1967). It is also the argument of 'established expectations' – that it is not unreasonable for people (and communities and governments) to form expectations on the basis of long-established practice, and to organize their lives and make important decisions on the expectation that these practices will continue in the foreseeable future. This is the argument advanced by David Hume (1962). It has considerable practical force, in that it recognizes the important point that EU arrangements cannot be constructed *de novo*, as if we were starting from nowhere with a completely blank sheet. The fact is that EU rules were not arrived at in a vacuum, but in the context of countries' long-existing fishing practices. So the EU could not operate as if behind the Rawlsian veil of ignorance (Rawls, 1972) – that would have done violence to established expectations – and this danger was a legitimate factor to have been borne in mind. However, doesn't such a justification amount to an assertion that whatever has occurred in the past is right? Moreover, how should these rights be transmitted from one generation of fishers to another? Merely because X has fished for a long time in any area may entitle X to exclusive fishing rights in that area, but when X dies, who is entitled to these exclusive fishing rights? Does X have the right to decide this? If so, on what grounds? Established expectation seems stretched rather far here.

Whatever the merits or otherwise of these arguments purporting to justify the entitlements generated by historic catches, it is important to note what sorts of entitlements are ruled out by the historic catches criterion. It rules out an idea of entitlement based on a crude form of egalitarian treatment of all member states – that is an equal amount of quota for each fish stock in each area allocated to each member state. Such an allocation would, of course, be absurdly unfair, giving land-locked states like Luxembourg and Austria equal shares to the UK, Ireland, France and Spain, who have extensive fishing interests. An alternative rule of entitlement could be an allocation of quota proportionate to the populations of the member states, but this would be almost as arbitrary and unfair, since population size has very little relevance to the fishing industries of member states. More relevant considerations would be the size of the country's population engaged in fishing, or the size of the country's fishing fleet, or the importance of fishing to the country's economy, and, indeed, such considerations are explicit in the notion of the Hague Preferences, which I discuss below under the concept of need. What I want to consider here is another idea of entitlement – that based on differential national property rights in member states' shares of EU waters. Such an idea of entitlement is often advanced by some member states (principally the UK and Ireland) when complaining about the low level of their share of the TACs. For example, the Irish South and West Fishermen's Organization (ISWFO) criticized the Irish government for failing to address this issue in its Presidency of the EU: 'Despite appeals from the industry, the government has refused so far to address the basic injustices to Ireland caused by the CFP . . . there is no mention of a revision of the deal that gives Ireland 5 per cent of the quotas in mainly low value fish, even though we have 16 per cent of EU waters' ('Presidency Under Fire', *Fishing News*, 16 August 1996, p. 2). Irish fishermen also point out that since Ireland only has 2.5 per cent of the EU fleet, they can hardly be held responsible for over-fishing, and therefore should not be subject to further cuts in quotas. Similarly, the UK has claimed that since it has 60 per cent of EU waters, it should be entitled to 60 per cent of the quota of fish allocated under the EU's CFP, instead of the 27 per cent given to it:

> The UK based its opening claim on its contribution to the resources, of the order of 60 per cent. This was totally unrealistic in terms of its historic share of the catches . . . Other Member States found such a high claim unacceptable . . . it was totally contrary to the Community principle that everything which goes into the Community pool is allocated between Member States on the basis of their needs, not on the basis of 'getting out what you put in'.   (Holden, 1994: 49)

Although Holden is mistaken in attributing to the Community a commitment to the concept of need as an overriding principle (the truth is that, as we shall see, the EU regards need as a tertiary principle, qualifying the operations of the primary principle of equal access, and the secondary principle of relative stability), he is right to draw attention to the fact that relative stability categorically rejects the idea that 'national ownership' of waters is the basis of allocation of quotas of fish to be caught in these waters.

A variation on this theme of entitlement through national ownership of resources is the claim, again put forward by the UK and Ireland, that since by far the majority (75 per cent) of fish caught in the EU are caught in the waters surrounding the UK and Ireland, so the majority of the fish quotas allocated by the EU should be awarded to the UK and Ireland. But on the principle of relative stability, such a claim is also rejected. So far as the CFP is concerned, both the waters of the EU member states and the fish stocks in these waters come under the jurisdiction of the EU, and whether or not member states claim ownership or sovereignty over either these waters or the fish stocks within them is neither here nor there so far as allocations of quota are concerned. For the CFP, entitlement is related to labour mixing by fishers, not to claims to sovereignty, territory or natural resources by member states.

It is at first sight surprising, therefore, to read this statement from the CFP Review Group: 'The focus of relative stability is inescapably national . . . without it, a scarce and finite resource could not be apportioned equitably . . . In combination, relative stability and the derogations from equal access demonstrate the CFP's provision for what are essentially protected national assets' (CFPRG, 1996, II: 9; I: 11). But what the CFPRG means is that since fishers belong to nations, when they build up track records of fish catching, these track records are attributed to the nations of which they are citizens; 'Relative stability . . . reflects actual fishing patterns (or national "track records") in the 1970s' (CFPRG, 1996, I: 11). In this sense, the principle of relative stability does respect nations' rights, but the 'national' element here is purely incidental, contingent and derivative; the claim is essentially founded upon individual fishermen's efforts, not national sovereign ownership of waters or fish stocks.

Assuming that this analysis is sound, is the CFP justified in prioritizing a labour-mixing criterion of entitlement over a national sovereignty criterion of entitlement? In my view, yes: of the two entitlement claims, although both are deeply flawed, the labour-mixing claim is superior to the national sovereignty claim. The labour-mixing claim at least has the moral weight that effort has been expended, and energy, initiative and investment have been used to achieve the historic fishing catch record, and this is a morally relevant reason for awarding preferential rights to future fishing opportunities. By contrast, the claim of national sovereignty at least to areas beyond the 12-mile coastal zone (the 12-mile zone could be justified as a security buffer) seems morally weak, resting on little more than accidental proximity. But both claims face formidable objections from advocates of a less discriminatory allocative system.

Need

Let us now turn to the concept of need. As noted above, the principle of relative stability contains within it two concepts of substantive justice: entitlement and need. What does the concept of need entail, and how does it relate to the concept of entitlement? The concept of need is explicit in the Hague Preferences – that is, the preferential treatment offered to fisheries-dependent communities in order to protect their 'vital needs'. How justifiable is this special treatment? One immediate criticism of the idea of 'vital needs' is advanced by free marketeers – that fisheries-dependent communities do not need artificial subsidizing; on the contrary, they need the discipline of competition so that they can make themselves economically strong, perhaps in another industry. The last thing they 'need' is to be lulled into a false sense of security; subsidies, in the form of privileged or preferential treatment, cannot be permanent.

In the discussion above on the equal access principle, we have already touched on another criticism of the argument for special treatment for fisheries-dependent communities – 'that all fishermen are "dependent" on fishing' (Wise, 1984: 167). To sustain a defence of the Hague Preferences, it would have to be demonstrated that the degree of fisheries dependence in the fisheries communities of the whole of Ireland, of Scotland and North East England was greater than the degree of dependence in the fishing communities of, for example, Brittany and Northern Spain. With over 22 per cent unemployment in Spain such a case may not be easy to make out. As Wise says, the concept of 'dependence' is politically loaded:

> ... much is made, in international fishery disputes, of how 'dependent' a country – or region – is upon fishing. However, such notions of 'dependence' defy precise objective definition ... in the process of political decision it is the 'importance' governments attach to the demands of fishing interests that is crucial,

rather than some elusive measure of national or regional dependence on fishing.   (Wise, 1984: 21, 24)

Moreover, even if, on the need criterion, it was justifiable to give preferential treatment to Irish and UK fishers over French and Spanish fishers, that necessarily entails less consideration given to entitlement rights based on historic catches. This signifies that the criterion of need overrides or trumps the criterion of entitlement in certain circumstances. Is this fair?

In a less explicit way, quota hopping also raises the issue of the tension between need and entitlement, according to its UK and Eire critics. As we saw earlier, there is nothing procedurally unjust about quota hopping. Indeed, it is a normal consequence of the freedom of capital guaranteed under the EU single market legislation that a national of one member state can set up in business in any other member state. It is also quite consistent with the entitlement principle: according to (Nozickian) entitlement theory, quota hopping would not violate rights provided there was no coercion entailed, since any transaction carried out in a just (that is, uncoercive) way from a just initial distribution (of historic catches) must itself be just, whatever its wider social consequences. After all, no boat owner is *forced* to sell their vessel (with licence and quota) to a foreign buyer. Moreover, quota hopping, like the equal access principle, is consistent with the EU's emphasis upon a single market and the rule of non-discrimination between citizens of member states.

However, many critics, especially in the UK, have argued that quota hopping completely undermines the intention of the principle of relative stability, which is to protect national needs. According to such a nationalistic interpretation of the relative stability principle (advanced by John Major, Tony Baldry and most UK fishermen's organizations (FOs)) quotas are allocated to nations for the benefit of indigent fisheries and any development which cuts the links between a quota and the nation to which it has been allocated flies in the face of the whole rationale for allocating quotas. Even the CFPRG claims that it 'clearly detracts from the supposed national focus of relative stability' (CFPRG, 1996, II: 12). On this interpretation, entitlement is not an end in itself (fairness or justice) but a means to another end – national need. So whenever the two concepts conflict, need overrides entitlement.

Holden agrees with these critics that quota hopping undermines the relative stability principle, but unlike them he regards this as a good reason for abandoning relative stability, not quota hopping.

> Those fishermen in whose Member States the 'foreign' vessels are registered object strongly to what they consider their quotas being 'stolen' from them and the issue generates much heated political controversy. However, the situation is now an established feature of the way in which the fishing industry operates and is in conformity with the fundamental political objectives of the Community, that of the single market. The situation is no different to companies establishing factories in member states other than their own. As relative stability can no longer achieve its intended political objective, there is no practical point in retaining it.
>
> (Holden, 1994: 241)

Although I have reservations about the nationalistic interpretation of the relative stability principle (for the reasons stated above), in so far as that principle does work to protect the needs of national fishing industries, it is threatened by quota hopping. Whether we choose to abandon the relative stability principle or to forbid quota hopping depends on the relative weight we place on the values of national need and equal freedom. However, on a practical level, if quota hopping is to be stopped, it can only be done by an unlikely decision of the EU to amend the Maastricht Treaty and downgrade its commitment to market freedom within the EU. As the CFPRG notes; 'Quota hopping cannot be addressed without considering these fundamental Treaty principles' (CFPRG, 1996, II: 12).

Conclusion on the principle of relative stability

The principle of relative stability does not seem procedurally fair, at least in spirit, since it systematically discriminates against new member states and ossifies 20-year old catching records. Nor does the principle of relative stability fully meet the tests of substantive justice, since its element of entitlement and its element of need are both flawed.

CONCLUSION

The above discussion has revealed considerable doubt about the fairness of the CFP. One response to that doubt would be the cynical conclusion that the CFP is not about fairness but about economic and political considerations, and that considerations of equity are well down the list of priorities of CFP policy-makers. A more positive response would be to point out that the key to the problem lies in identifying the different categories of persons or groups to whom the CFP should be fair. Should the CFP be fair to (1) individual fishers? (2) local fishing communities? (3) national fishing industries? (4) member states? (5) European Union fish consumers? (6) fish species? (7) marine ecosystems? In this chapter I have been concerned only with categories (1), (2), (3) and (4), ignoring issues of fairness to consumers, fish species and marine ecosystems. Confining our attention to these first four categories, we can see that different conclusions on the issue of fairness of the CFP may be reached according to the different perspectives of the constituencies from which that issue is viewed. For instance, quota hopping would be viewed as fair by individual fishers and by some member states, but not by fishing communities, national fishing industries or other member states. Similarly, the entitlement element (historic catch records) of the principle of relative stability would be regarded as fair by most individual fishers; the needs element would be regarded as fair by local fishing communities; and the compensation element would be regarded as fair by national fishing industries. In other words, there may be some elements of fairness in the CFP that each affected constituency can acknowledge. A third response would be to explain that an apparent breach of fairness by the CFP should be understood in the light of the wider context of rules of fairness operating between member states within the EU. For example, dilution of the derogations on equal access in order to allow Spain and Portugal a greater share of EU fish stocks may be part of a complex balancing act in which fishing interests are traded off against other national interests, with a view to reaching an overall fair outcome. Such political deals may not be devoid of fairness, but may incorporate within them a broader sense of fairness than that which is confined to fishing interests alone. Whatever one's opinion on these different perceptions and levels of fairness, it is clear that the issue of the fairness of the CFP is much more complex than is generally acknowledged.

NOTES

1. I am grateful to Peter Jones, Derek Bell, Mark Bevir, Simon Caney, Ella Ritchie and Tony Zito for their helpful comments on earlier versions of this chapter.
2. The terms 'fishers' is used instead of 'fishermen' since some fishers are female.

BIBLIOGRAPHY

Burke, E. (1967) *Reflections on the Revolution in France* (London: Dent).
CFPRG (1996) *A Review of the Common Fisheries Policy Prepared for UK Fisheries Ministers by the CFP Review Group*, 2 vols (London: MAFF).
Deas, B. (1996) 'Coastal State Management' (this volume).
Holden, M. (1994) *The Common Fisheries Policy: Origin, Evaluation and Future* (Oxford: Fishing News Books).
Hume, D. (1962) *A Treatise of Human Nature*, Book III, Part II (London: Fontana).
Locke, J. (1962) *Of Civil Government* (London: Dent).
Nozick, R. (1974) *Anarchy, State and Utopia* (New York: Basic Books).
Rawls, J. (1972) *A Theory of Justice* (Oxford: Oxford University Press).
Symes, D. and Crean, K. (1995) 'Historic Prejudice and Invisible Boundaries: Dilemmas for the Development of the Common Fisheries Policy', in Blake, G. H., Hildesly, W. J., Pratt, M. A., Ridley, R. J. and Schofield, C. H. (eds), *The Peaceful Management of Transboundary Resources* (London: Graham & Trotman), pp. 395–411.
Wise, M. (1984) *The Common Fisheries Policy of the European Community* (London: Methuen).

Index